Microservices in SAP HANA XSA

A Guide to REST APIs Using Node.js

Sergio Guerrero

Microservices in SAP HANA XSA: A Guide to REST APIs Using Node.js

Sergio Guerrero
Texas, United States

ISBN-13 (pbk): 978-1-4842-6117-0   ISBN-13 (electronic): 978-1-4842-6118-7
https://doi.org/10.1007/978-1-4842-6118-7

Managing Director, Apress Media LLC: Welmoed Spahr
Acquisitions Editor: Divya Modi
Development Editor: Laura Berendson
Coordinating Editor: Divya Modi

Cover designed by eStudioCalamar

Cover image designed by Pixabay

Distributed to the book trade worldwide by Springer Science+Business Media New York, 233 Spring Street, 6th Floor, New York, NY 10013. Phone 1-800-SPRINGER, fax (201) 348-4505, e-mail orders-ny@springer-sbm.com, or visit www.springeronline.com. Apress Media, LLC is a California LLC and the sole member (owner) is Springer Science + Business Media Finance Inc (SSBM Finance Inc). SSBM Finance Inc is a **Delaware** corporation.

For information on translations, please e-mail booktranslations@springernature.com; for reprint, paperback, or audio rights, please e-mail bookpermissions@springernature.com.

Apress titles may be purchased in bulk for academic, corporate, or promotional use. eBook versions and licenses are also available for most titles. For more information, reference our Print and eBook Bulk Sales web page at http://www.apress.com/bulk-sales.

Any source code or other supplementary material referenced by the author in this book is available to readers on GitHub via the book's product page, located at www.apress.com/978-1-4842-6117-0. For more detailed information, please visit http://www.apress.com/source-code.

Printed on acid-free paper

*To my wife, Amy, who always supports my goals
and achievements. Thank you for being there in my journey.
I would not be where I am without you.*

To my children, Oliver and Emma.

*To my late parents Jorge and Maria del Rosario
and my late brother Francisco.*

Table of Contents

About the Author

Sergio Guerrero is a seasoned architect who specializes in custom application development and systems integration. With over 10 years in software development, he has designed and built solutions for several clients in various industries (industrial, defense, retail). He is an active participant in the SAP community in English and Spanish channels. Sergio has spoken at several tech conferences such as SAP TechEd, SAP Sapphire, and SAP Inside Track in Mexico and Colombia.

About the Technical Reviewers

Attaphon Predaboon is a senior data and analytics consultant working for EY with multiple SAP HANA and BW certifications. He has led the teams in proofing of concepts and deployments of SAP HANA 2.0 XSA for a large-scale global implementation. He has worked with architecture and implementation of SAP BW, HANA, and big data technologies.

Murlli Maraati is a SAP analytics architect with a background in SAP BW, SAP HANA, SAP Cloud Platform, and SAP Analytics Cloud. With over 7 years of SAP experience, Murlli has been involved in multiple support and implementation projects. He is experienced in SAP HANA 2.0 (XSA/Web IDE), HANA Security, Node.js, REST API, Virtual Data Model (VDM) with Embedded Analytics using calculation views and CDS views in S/4, and other custom SAP applications.

Acknowledgments

I am thankful to my wife, Amy, for her support and encouragement while writing this book and in my personal life above all circumstances. Thank you for listening to me ramble sometimes. I am very humbled and grateful for the Apress team, Divya and Laura, for reaching out, and the technical reviewers, Attaphon Predaboon and Murlli Maraati, for providing a different perspective while writing this book, for their support, and for guiding me through the journey of this professional achievement.

Introduction

Microservices are being developed to address application issues that businesses face today, such as monolithic architectures, being able to develop and deploy software between different platforms, and being able to quickly respond to business events. It is now possible to develop node microservices with the introduction of the advanced architecture in SAP HANA. This book is intended to show the development cycle, debugging, and deployment of Node JS microservices in the SAP HANA platform. Some tools and ways of utilizing them are showcased in the interest of helping the reader to feel comfortable in developing their next enterprise-ready microservice/REST API in SAP HANA. At the end of the book, the reader will have completed building a REST API using commonly used HTTP methods such as GET, POST, and DELETE while using data models exposed as calculation views to read data or by calling stored procedures to update, insert, or delete records.

Chapter 1 provides an overview of the SAP HANA architecture and how its advanced architecture is based on Cloud Foundry concepts.

Chapters 2–4 start with the security model to understand how security is implemented following the Cloud Foundry architecture and how it is utilized during the book exercise. The book continues with building database objects such as tables, calculation views, and stored procedures. Then, the REST API is built using the Node JS language. All API endpoints are unit tested via POSTMAN (free tool). Debugging occurs from the Web IDE on Node JS code and on the stored procedures to debug SQL.

Chapter 5 concludes the book explaining software versioning, deployments from the SAP Web IDE, and scenarios to scale the microservices from the HANA Cockpit for XSA development.

All source code is available on GitHub via the book's product page, located at www.apress.com/978-1-4842-6117-0. For more detailed information, please visit www.apress.com/source-code.

CHAPTER 1

Architecture of SAP HANA XSA

Welcome to how to build microservices in SAP HANA XSA using Node JS. In this chapter, I explain the SAP HANA architecture, the evolution from HANA 1 to HANA XSA, the Cloud Foundry (CF) architecture, and also how open source plays a very important role in the development of microservices in HANA XSA.

SAP HANA

Enterprises have ERP (enterprise resource planning) systems where they run business processes autonomously or by utilizing employees working at these companies. ERP systems such as SAP have provided proven technologies that utilize System Applications and Products (SAP) in data processing to analyze, plan, and produce goods and services. These goods and services expand across industries and sometimes have global presence.

SAP HANA (Figure) is an in-memory platform containing a database, a web server, and other engines that make up a highly available, highly performant, analytic engine. SAP HANA as a database is primarily columnar oriented in nature and its in-memory capabilities combine OLAP and OLTP in the same system. These and other features that will be shown in this book make this system a very popular one in the enterprise

© Sergio Guerrero 2020
S. Guerrero, *Microservices in SAP HANA XSA*, https://doi.org/10.1007/978-1-4842-6118-7_1

around the world. As we know, OLAP is used for large volumes of data in traditional database management systems. OLTP is used for data warehousing where a large number of short transactions are utilized. SAP HANA has both capabilities.

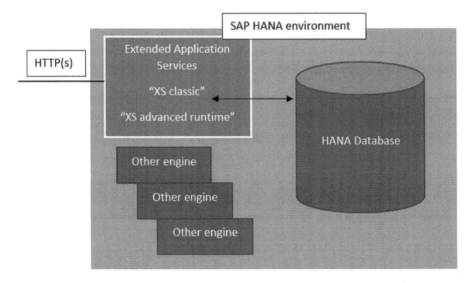

Figure 1-1. *SAP HANA environment displaying the HANA database and the XS engine*

Understanding ERP systems and having in-memory platforms such as SAP HANA have become crucial because business processes need to make near real-time decisions responding to demands and new trends. Business scenarios such as those involving manufacturing, planning, scheduling, retail, energy, and industrial production, among many others, require these capabilities to innovate and reimagine their possibilities on daily operations. Can you imagine a business where all its data and processes could be fully integrated without having to do some business operations in one system and other operations in another system? That type of process would be inefficient, and many redundant copies of data would add chaos to the entire landscape.

As of today, SAP HANA has those capabilities. SAP HANA can serve as a database. Within the same system, there are various ways to bring data together in order to create business data models. One of the many ways to integrate data from other parts of an enterprise is known as SLT (SAP Landscape and Transformation).

SLT is an ETL (Extract, Transform, Load), trigger-based, replication technology from SAP ECC into a SAP HANA schema. SLT can also be integrated between other database systems such as Microsoft SQL Server, Oracle, and SAP HANA as shown in Figure 1-2. SLT can be configured from the SAP GUI, or once a system connection is established in SAP HANA Studio, a developer can set the replication of tables (as full, partial, stop, resume, or suspend) using the data provisioning link from the quick view menu in SAP HANA Studio.

Why is this technology important? Because we will no longer need to separately query external data sources to analyze them and then aggregate them. Once the external data source tables change, their triggers will automatically replicate the latest changes into SAP HANA, and then additional business data models can be created within SAP HANA in order to provide a full understanding from logically separated systems. Check out SAP's SLT official documentation to get a deeper understanding of this technology (`www.sap.com/products/landscape-replication-server.html`).

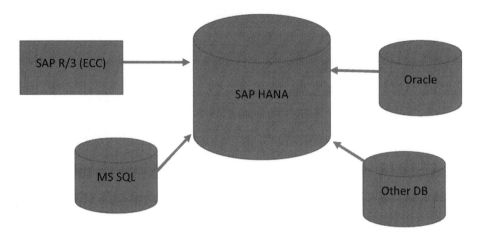

Figure 1-2. *Data integration from external data sources into*
SAP HANA

Another way to integrate data models from and into SAP HANA, ECC, and external database systems is via smart data integration, also known as SDI. SDI is a technology that makes connections to external systems and makes virtual copies, in batch or real time, of the data structures on those systems (Microsoft SQL Server and Oracle, among others) using custom adapters as represented in Figure 1-3. SDI can be set up from SAP HANA Studio where a developer can provision virtual tables and eventually create integrated business data models with those virtual tables. In SAP HANA XSA, SDI can be configured from the Database Explorer.

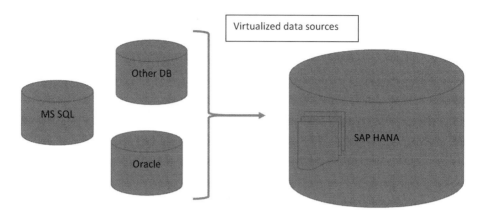

Figure 1-3. *Virtualized data sources*

Another available feature within SDI is that developers can create flowgraphs (Figure 1-4) as workflows from the SAP Web IDE. These tasks facilitate the creation of replicating data or performing additional tasks without having to write actual code, other than using the drag and drop feature to map out the flow of data from source to target. In some cases, additional ETL can be achieved by using the aggregation, join, union, or pivot node operators in these workflows. Once a workflow is created, then it becomes a task that can be scheduled as shown in Figure 1-5.

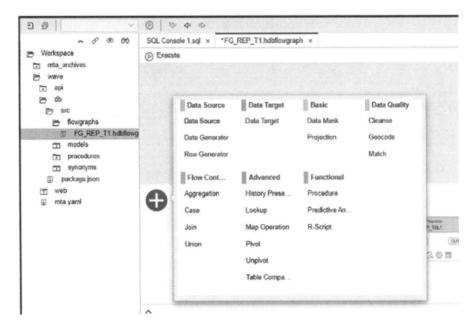

Figure 1-4. SDI flowgraph operations

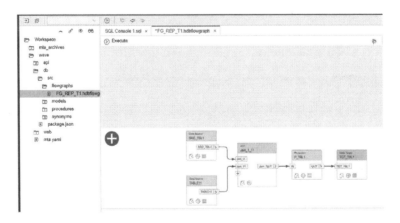

Figure 1-5. SDI flowgraph nodes

To run this task, do so by querying it from the SQL console with the following syntax:

START TASK <TASK_NAME> as shown in Figure 1-6.

To schedule this workflow, do so by creating an XS Job (.xsjob) and configure its job details, such as frequency, interval, time of execution, the object to be run, and any parameters to be passed to the flowgraph.

Figure 1-6. *Starting a TASK from the SQL console*

As we transition from the database side of SAP HANA and into the web application front end, SAP HANA provides a web server called Extended Applications Services, known as the XS engine, that allows the SAP HANA platform to expose and ingest data via HTTP(s), send emails, execute chronicle-like jobs, and also make outbound external web request calls, for example, calling a third-party API (application programming interface). The XS engine is a relatively small footprint application server, web server, and development environment. In SAP HANA 1.x and HANA 2.x, the XS engine also has its own server-side JavaScript language called XS JavaScript (XSJS). XSJS was created by SAP to provide HTTP access to data and interact with the SAP HANA database. The XSJS language was created on top of the SpiderMonkey Firefox engine. This language has several internal APIs such as the ones listed in Table 1-1.

Table 1-1. *SAP HANA XSJS APIs*

Description	$ API	Used for
Database	$.hdb	Integration to the HANA database from XSJS
Legacy database	$.db	Integration to the HANA DB (deprecated)
HTTP outbound	$.net	Making outbound calls (external to XS engine)
Request (incoming)	$.request	Analyzing a request to the XS engine
Job scheduling	$.jobs	Creating and maintaining XS jobs
Security	$.security	Check data with antivirus and also store secure data
Trace	$.trace	Set up log traces while executing XS code
Text	$.text	Text analysis and text mining * require additional licensing
Util	$.util	Zip and xml parsing

While these XS APIs are still working on HANA 2 under the XS compatibility mode, there is also another way to interact with the SAP HANA database in SAP HANA XS advanced. In Chapter 4, there will be equivalent approaches using a Node JS implementation and @sap node modules.

The XS engine became popular for enterprises because it allows near real-time in-memory processes within SAP HANA, and it allows companies to interact with their ERP data in an easier, industry standard way called REST APIs. REST stands for Representational State Transfer, which in layman's terms means that the state of a system is derived from data and not from a session state. This is also true for the Internet which is stateless. API allows external consumers of a system to interact with it via Internet protocols without having to provide direct access to a back-end system. The consumer of the API needs to know the back-end API URL, the data structure to send to the API, and what they will receive as a response (if anything). Some companies wanting to minimize the technology stack footprint started moving away from external technologies like Microsoft's .NET or Java and created XS applications, while others still use these external technologies to consume SAP HANA back-end services from the XS engine.

An additional benefit of creating XS applications in SAP HANA is that an application can be as simple as the database in the back end or include a web service (REST API) or even a front-end HTML5 modern application as shown in Figure 1-7; otherwise, in a full stack scenario, it can contain all of these components, as decoupled microservice applications. REST APIs act as a black box from the consumer of the service – consumers should not care what processes are run in the back end. For their own sake, they just need to know the URL, the contract, and if the service will return a response so they can handle it.

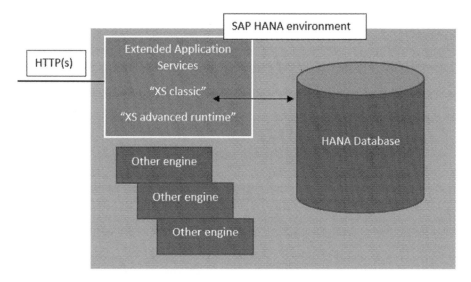

Figure 1-7. *SAP HANA API request via the XS engine*

Cloud Foundry (CF)

Cloud Foundry (Figure 1-8) is an open source cloud application platform. Cloud Foundry is made up of a group of companies such as IBM, SAP, Microsoft, Google, and others, coming together to set up industry standards allowing applications to be runnable across cloud providers so that developers can focus on application development and not on creating the cloud. One of the benefits of creating applications that are Cloud Foundry ready is that these applications are essentially decoupled from the infrastructure so when they are ready to be moved to on-premise, public, or other private clouds, these applications are easily migrated from one environment/cloud to another. Before getting into applications, it is appropriate to look at Cloud Foundry's hierarchical organization and some important concepts we will be interacting with: organization, space, application, services, and roles as shown in Figure 1-8.

1) An **organization** (org) is defined as a development account where one or more developers (known as contributors) have access known as user accounts. These contributors have special roles assigned to them in order to manage, audit, and build cloud-ready applications or grant access to others. For demonstration purposes, the examples provided in this book will have an organization already created, and it is in an active status. Only an admin can change the status of this org; however, we will assume it is active for the rest of our exercises.

2) Following an organization, there are one or more spaces. A **space** is defined as a partition within the org dedicated to build, deploy, and maintain applications. Every application is scoped to one space. Having an application will eventually require granting roles for

access to that application. Application roles belong to one and only one space. Later, in the microservice REST API section, it will be necessary to set our application to be assigned to a space.

3) **Applications** and services are the next level within the hierarchy of CF. These applications and services allow users to see some data or consume data.

4) Users may have one or more roles (role collections are templates) assigned to them. These roles allow the users to develop, access, manage, or audit applications, spaces, or even orgs. Application security is a very complex part of software systems. Security in the cloud (and in HANA XSA) can be its very own extensive topic. There are several other books currently available that explain SAP HANA XSA security in extensive detail. For this book, there will be enough security to access a table and a view and execute some stored procedures within a SAP HANA XSA environment.

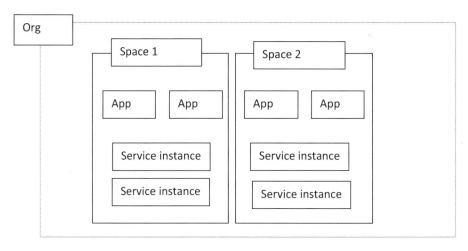

Figure 1-8. *Cloud Foundry organization hierarchy*

Applications that are infrastructure agnostic are more dynamic in nature and can easily be deployed in minutes without having to make application changes. This type of architecture is known as container-based architecture. CF is container based and it allows applications to be developed and run in any programming language (BYOL – bring your own language) and framework of choice.

As mentioned earlier, Cloud Foundry decouples the application from the infrastructure. Additionally, the Cloud Foundry resources (applications/services) are accessible via a route and an authorization service called UAA (User Account and Authentication service) as shown in Figure 1-9.

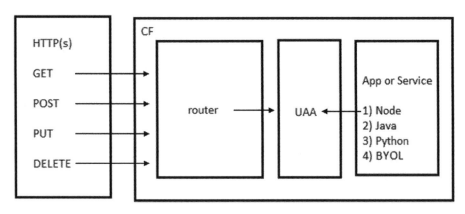

Figure 1-9. *REST API following CF router and UAA service*

The main role of the UAA service is to issue client tokens for authorization to authenticated requests in the form of OAuth tokens. These OAuth tokens come in the form of JSON and are generated with a limited life span. Once the tokens expire, a different request to get a token must be sent and authenticated, and the UAA service needs to generate a new token for the subsequent requests. Similar steps will be demonstrated later in the book to show the flow of making a request on a SAP HANA XSA microservice REST API. Keep in mind the CF concepts are very important to understand the implementation of SAP HANA XS advanced architecture – more details on the "SAP HANA to HANA XSA" section.

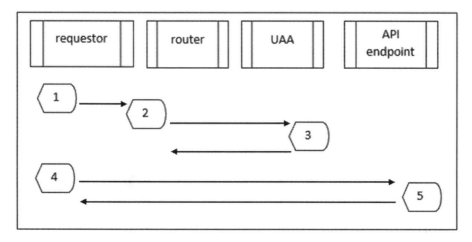

Figure 1-10. *CF client request to endpoint steps*

Steps on the client-to-back-end resource endpoint via UAA (Figure 1-10) are as follows:

1) The requestor makes a request of some URI in our (cloud) environment.

2) The router sees and recognizes the request and redirects to the UAA service for authentication.

3) The UAA receives the request and

   a. Authenticates and returns an OAuth token

   b. Denies and returns unauthorized

4) The requestor must send the desired request along with the UAA token so that the back-end resource can recognize the authenticity of the request. Sometimes this is referred to as bearer token.

5) If the token is valid, the back-end resource returns
the response to the requestor; otherwise, it will
return an unauthorized error.

There are different ways to interact with the Cloud Foundry instances.
One of them is via the command-line interface (cf CLI). The CLI can be
combined with Linux commands, such as pipe, grep, and others. There
are various commands in which we will have to interact with the cloud/
XSA architecture, such as logging in, setting up services, creating, editing,
building, deploying, or even scaling an application or service. The Cloud
Foundry CLI is a versatile tool that allows developers to interact with
their cloud instance. It is easy to automate and script other processes that
otherwise would require a human to type commands. It minimizes the
risk of introducing errors while interacting with the cloud and keeping it
available for developers and admins who would require knowing cloud
container details.

Examples of such commands are

cf **login** <USER_NAME> # to log in to the cloud
instance

cf **apps** | grep <APP_NAME> # to retrieve an
application from the list of applications

cf **push** <APP_NAME> # pushing an application to
the cf instance

cf **scale** <APP_NAME> -m <MEMORY_SIZE> -i
<INSTANCES>

A comprehensive list of available commands is shown here: `http://`
`cli.cloudfoundry.org/en-US/cf/`.

Otherwise, once Cloud Foundry CLI is installed on a system, one can
run the command cf --help to see how to utilize a command and to see
what flags it has as shown in Figure 1-11.

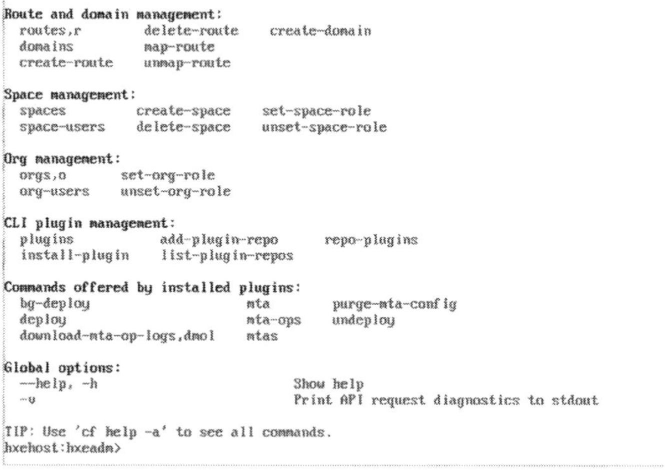

```
Route and domain management:
  routes,r          delete-route     create-domain
  domains           map-route
  create-route      unmap-route

Space management:
  spaces            create-space     set-space-role
  space-users       delete-space     unset-space-role

Org management:
  orgs,o            set-org-role
  org-users         unset-org-role

CLI plugin management:
  plugins           add-plugin-repo       repo-plugins
  install-plugin    list-plugin-repos

Commands offered by installed plugins:
  bg-deploy                    mta          purge-mta-config
  deploy                       mta-ops      undeploy
  download-mta-op-logs,dmol    mtas

Global options:
  --help, -h                          Show help
  -v                                  Print API request diagnostics to stdout

TIP: Use 'cf help -a' to see all commands.
hxehost:hxeadm>
```

Figure 1-11. *cf command-line interface (cf CLI)*

With such a dynamic environment comes the need to leverage additional open source software. Many companies are slowly adopting the concept of developing and reusing open source software to benefit from other contributors' development and avoid having to re-create repeatable software solutions. There are several open source software package managers that currently exist and are very popular among the developer community; one of them is Node Package Manager (npm).

Before getting into npm, it is important to understand Node JS. Node JS is an asynchronous, dynamic, nonblocking, event-driven, server-side, JavaScript language built on Google Chrome's V8 JavaScript engine. Unlike traditional JavaScript, it does not require a browser to run. It can run on a computer as an autonomous program if the Node JS runtime exists in such a machine. In addition to the runtime, there may be additional packages/modules that a developer would need to leverage in order to develop robust programs such as REST APIs. How do these other packages/modules get into our development? That is where the package manager comes in handy – assuming the developer has privileges on the machine to be able to download the package manager and be able to run and install additional packages. When creating a node program, there are some files that are important to highlight:

a) **The main Node JS file** – Let's call it index.js.

b) **package.json** – This file holds metadata about the node application like runtimes, dependencies from other modules, and some script commands that are helpful to run, test, and deploy our node application. This file is used when invoking the package manager in order to try to download and integrate specific dependencies and versions of those dependencies into our application.

A simple "Hello World" program running from a text editor looks like Figure 1-12.

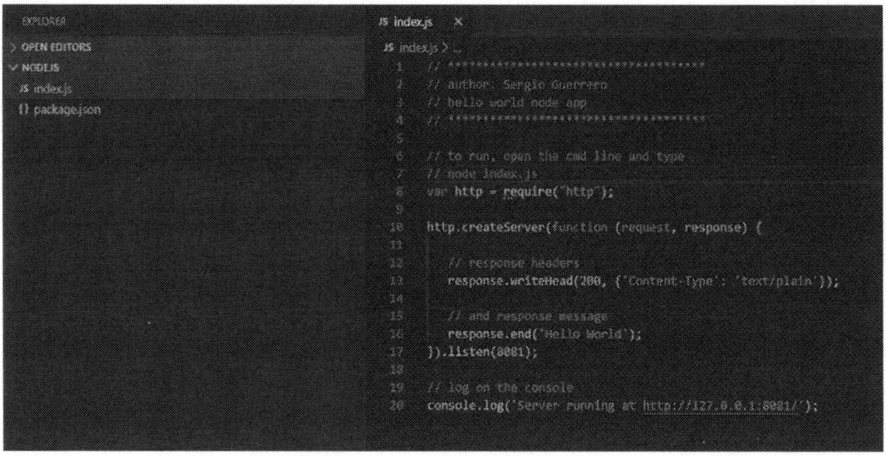

Figure 1-12. *Node JS simple "Hello World"*

And when running on the browser (since a server was created to serve a response using the HTTP node module), as shown in Figure 1-13.

Figure 1-13. *Output of Hello World from the browser*

Imagine text can be served, following the logical next use case would be to be able to serve HTML as a web application or even create a REST API back-end service to serve and receive requests; this will be shown in a later chapter.

The Node Package Manager (npm) is an online repository of open source software packages/modules. NPM is free of cost for developers making public software (check the npm pricing section for private development licensing cost). Developers can utilize these node modules to build and extend their applications. When navigating the Node Package Manager, www.npmjs.com/, one can find details such as version, dependencies (of other modules), history of downloads, number of downloads, license, repository URL, collaborators, author information, as well as other important details for the open source module. Some of those modules are extensive enough to become frameworks themselves. One popular module/framework is called Express. Express is a very powerful, minimalist framework to build REST APIs. The Express framework has great features of robust routing, focusing on fast performance, super-high test coverage, and content negotiation, and it also has features to generate applications quickly. There are free modules for any type of work, process, or task, for example, the Passport module to authenticate requests, Socket.IO for real-time two-way event-based communication, and PDF generators, just to mention a few. Some of these modules will be used later in this book to build a Node JS microservice including routing, authentication, and request/response handling.

What are microservices? Microservices are decentralized architecture in which programs such as front-end, back-end, and database applications can work, scale (grow/shrink), build, deploy, start, and stop independently from others within an environment. Microservice architectures are modern ways to build scalable, cloud-ready, robust solutions that are loosely coupled, independently deployable, and primarily organized around business capabilities and owned by a small team as represented in Figure 1-14.

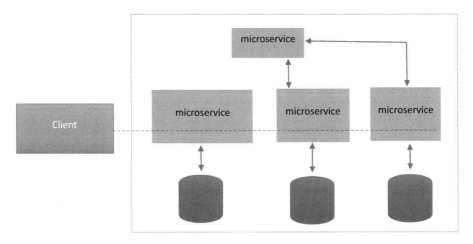

Figure 1-14. *Microservices*

Open source

Open source software has become very critical for developer teams to quickly integrate unknown ways to accomplish certain tasks with some of the software that has been developed and shared via package managers such as the Node Package Manager.

Just like anything else, there are benefits and drawbacks to utilizing open source software. Some of the benefits and drawbacks of utilizing open source software are shown in Table 1-2.

Table 1-2. *Benefits and drawbacks of using open source*

Benefit	Drawback
There are many open source libraries free of use	Most comes from unknown publishers who are open source contributors
These libraries usually have code samples and plan documentation easy to understand and follow	It may take time to understand license details
Get integrated in almost no time (usually minutes)	Developers must keep up with library versions and updates
Package manager shows the latest version and the number of downloads which shows the popularity of the module	Software is for use as is and no warranty
Speeds up development costs since developers do not need to reinvent common processes	Experienced developers usually understand the process of integration; however, less experienced developers require some level of explanation of how to use this type of software
Cost savings on companies needing to develop some common scenarios	Vulnerable to malicious users
Less expensive than commercial solutions	Lack of extensive support (use at own risk)

In addition to benefits and drawbacks, there are also company rules that may prohibit the use of certain open source software under certain licenses. Usually bigger enterprises have processes in place that require their legal departments to evaluate the need to use open source software, when this software has not been developed from within their resources. If your enterprise falls within these limitations or safeguards, please consult with your software architects or legal department on what requirement and open source libraries you would like to utilize in order to understand the possibility of your company being able to make use of such open source software.

Some of the most common open source licenses are (please read the respective license for any of the open source software you want to use)

1) Apache License

2) MIT License

3) GNU General Public License

4) Mozilla Public License

5) Eclipse Public License

SAP HANA to HANA XSA

Many companies are existing customers of SAP and utilize SAP HANA. These companies are the primary audience for learning how to migrate from SAP HANA classic to SAP HANA XSA (blue-green field deployment). Blue-green deployment is a technique to reduce downtime by running two identical production systems, where one is live (blue) and one is idle (green), deploying and performing full testing on the idle system and then switching traffic from the live (blue) to the idle (green) system and doing the same deployment and full testing on the new idle (blue) system. Those companies that are not going through an upgrade exercise will easily be able to start their journey of starting in SAP HANA XSA as they will not need to worry about existing development to be compatible in the newer environment (green field which is all new). Depending on the status of an upgrade path, it is recommended to have good planning in place to minimize or, better yet, eliminate downtime. In a later chapter, simulation of implementing a green field deployment approach will be explored since HANA 1 did not have support for the Node JS runtime* (as an XS classic landscape).

The first and most important step of going through this new implementation exercise is to be able to understand the Cloud Foundry principles. Let's take a moment to mention a few important concepts from Cloud Foundry and how they apply in SAP HANA XSA. These concepts are fundamental in understanding the architecture, security, and integration of web applications, microservices, and REST APIs.

From a hierarchical point of view, XSA, as described in Cloud Foundry, starts with an *organization*. In HANA XSA, an organization is also a grouping of spaces managed by an administrator. A space is a holder of zero or more applications and/or services (managed or user provided). A managed service is one provided by the environment. An example of this type of service in SAP is called XS UAA (the User Account and Authentication service). A user-provided service is a custom service (CUPS) created by a developer, and this type of service is usually not offered by the marketplace. An application is a group of one or more modules for a specific purpose such as an HTML5 application or a Node JS REST API. The applications are accessed by users containing roles with specific access, also known as scope.

Now that the hierarchical structure of Cloud Foundry has been explained and the same structure applies to SAP HANA XSA, it is important to begin planning the exercise that will be demonstrated throughout the book.

Prior to the demonstration, the differences in artifacts between the two architectures of SAP HANA and SAP HANA XSA are important to highlight. Objects will be affected in one or more of the following ways:

a)   Be unaffected (no change)

b)   Be changed (either upgraded or updated)

c)   Be deprecated (will not work going forward)

d)   New objects that did not exist before

For demonstration purposes, see the following for a very similar project as it existed in SAP HANA 1 and now needs to be re-created (or updated for SAP HANA XSA).

Assume this project has a multi-module project containing a set of database artifacts such as a set of tables, HANA views, and stored procedures, to mention a few. These database artifacts will work in conjunction with each other and with a layer of XS JavaScript REST API services to read, create, update, and delete data in SAP HANA system as shown in Table 1-3. These back-end services will be consumed from a front-end HTML5 application as shown in Table 1-4.

Table 1-3. *Database artifact differences between SAP HANA 1 and HANA 2 XSA*

Module	HANA 1 artifact	HANA 2 XSA artifact
Database	Schema	No change – created from the SQL console
	Schema table	Requires synonym
	DUMMY table	Requires synonym (similarly for cross-schema object access)
	Calculation view	No change
	Stored procedure	No change
	Table function	No change
	Role	No change
	~~Scripted calculation view~~	**Must** convert to table function or stored procedure *
	~~Analytical/Attribute views~~	**Must** convert to calculation views
	N/A	**hdbgrants** – new artifact, created from the SQL console
	N/A	**Role collection** – new artifact (group of roles)
	N/A	**hdi-config** – metadata for the HDI container
	N/A	**hdi-namespace** – metadata for the db module

Table 1-4. *XS artifact differences between HANA 1 XS and*
HANA 2 XSA

Module	HANA 1 artifact	HANA 2 XSA artifact
Old project files	~~xsapp~~	Does not exist in XSA
Node (backward compatible)	N/A	**xs-app.json** – new artifact that contains the security authentication types, routing, welcome file. Should a REST API need to be created, an HTML5 module must be created and it should contain routing. The API will use the URL of where the HTML5 module is deployed + the relative path to the node module**
	xsjs xsjslib xshttpdest	No change under the XSJS backward compatibility mode
	xsjob	No change
	N/A	(node) .js files and npm modules (public or @sap)
HTML5	View (xml/js)	No change
	Controller(js)	No change
	index.html	No change
MTA project	N/A	**xs-security.json** – new artifact that describes the application scopes. It is an optional file in case scope of the application must be enforced
	N/A	**mta.yaml** – new artifact that contains the module relationship and metadata for the project, UAA services

The difference between table function and stored procedure is that a table function cannot execute CUD (create, update, delete) statements, and they must return a table structure as their output. Stored procedures, however, do not need to return any type of output parameters after they are executed. Stored procedures are the only approach for these types of operations (create, update, delete).

Moving from the code into the next layer, one must think about the application lifecycle manager. Where is this code going to be stored, retrieved, and deployed to? In the SAP HANA 1 environment, there was the SAP HANA repository. One of the drawbacks of this approach is that multiple developers cannot work simultaneously on the same set of files/changes. In SAP HANA XS advanced architecture, the SAP HANA repository no longer exists; instead, SAP eliminated the idea of having their own repository solution and suggested companies to store their code in external code repositories such as GIT (Bitbucket or Microsoft TFS). There are many great benefits to this approach – things like introduction to DevOps, allowing multiple developers to work on code feature vs. bug fix, and most importantly, the beginning of CI (continuous integration) in SAP HANA.

For simplicity of this book, it is assumed that everyone following the exercises can connect, commit, and clone from a GIT repository using their own license – the book will showcase the use of an account with a free license.

In order to work with GIT, a developer must have access to this repository, be familiar with the GIT commands shown in Table 1-5, and either execute from the command line (if you use this approach, you will need to download the GIT CLI) or interact with them from the SAP Web IDE which already has an interface to a GIT repository as shown in Figure 1-15. The GIT commands are fundamental in the development and maintenance of modern software. It will be required to understand how each of these commands work and when to use them. Many more commands are available, but not all are shown here. Table 1-5 shows a list of the most often used (and not in order).

Table 1-5. *Git commands*

Command	Description	Example
Clone	Clones the external repository to a local workspace	git clone <url>
Commit	Submits changes from local workspace to the remote repository	git commit -m <comment for commit>
Stash	Saves the changes you do not want to commit immediately	git stash
Push	Pushes your changes to a remote (master) branch	git push <remote> <branch>
Pull	Pulls the changes from the remote branch	git pull <remote> <branch>
Rebase	Process of moving or combining a sequence of commits to a new base commit. The main purpose of a rebase is to maintain project history	git rebase <base>

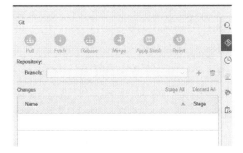

Figure 1-15. *Git pane in the SAP Web IDE*

One last thought to understand about an external repository, such as GIT, that may add some confusion to developers and is often explained incorrectly is to be able to understand the difference between Git, GitHub, and GitLab. What is the difference between Git, GitHub, and GitLab? GitLab and GitHub are both web-based Git repositories. Git is the mechanism to manage development projects and files during the software development lifecycle and stores these changes in a storage called repository. More than just a repository, git-like repositories are the beginning of CI/CD (continuous integration and continuous delivery) because of automated software builds of software via these commands by creating a flow.

A huge recommendation for new developers on this type of repository and software lifecycle management is to read the best practices on how to use them and understand procedures before starting projects. It will truly make a difference in the entire process. Some recommendations are

a) **Test your software before you commit** – Avoiding the introduction of bugs.

b) **Commit often** – Small and related commits allow easier code sharing.

c) **Write detailed commit messages** – Help others understand your committed code.

d) **Use branches** – To distinguish bug fix and new features from the main branch (golden copy).

e) **Do not panic** – Things may go wrong when you start but read about how to solve your problem.

There are many other best practices that will be shown later, but keeping these in mind before starting development will help developers maintain better software in the long run.

If external code repositories such as GitHub have not been used before, please start with a sample project. Create it, add a few files to the project, commit, add useful messages, and try making a branch and merging it back to the main line. The more prepared a developer is before the real project, the better they and their team will be when it comes to a real project.

Here are some steps to create a first-time sample project. Note that this is not the same project from the REST API exercise, and it is simply for illustration purposes – assuming there is a GitHub account already created and available for use as shown in Figure 1-16.

1)   Navigate to the ***Repositories*** tab and click **New**.

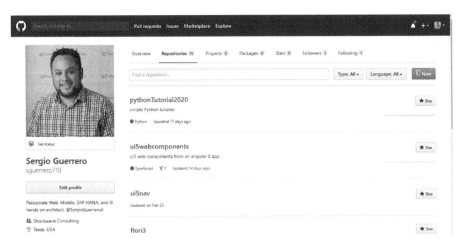

Figure 1-16. *GitHub account*

2)   Provide repo details such as name and description, select either private or public, and select any license to be included with this project repository as displayed in Figure 1-17. Click **Create repository**.

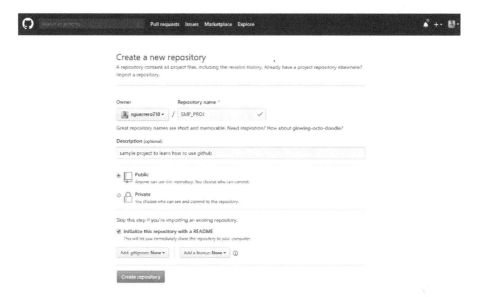

Figure 1-17. *Creating a GIT repository*

3) Once created, there will be a screen like the one in Figure 1-18. It is important to note what is shown on this screen. Take a moment to browse through the different tabs to see your Code, Issues, Pull requests, and all the tabs across. This screen has a lot of relevant information that can make or break development. It is very important to dedicate enough time here before proceeding. Additional documentation can be found on `https://help.github.com/en/github/getting-started-with-github/create-a-repo`. Notice that this screen will also allow developers to **Clone or download** a repository. The button for clone or download will provide a URI that can be used to set an external repo from a different tool such as the SAP Web IDE or any other IDE of the developer's choice that supports git integration such as Visual Studio Code as shown in Figure 1-18.

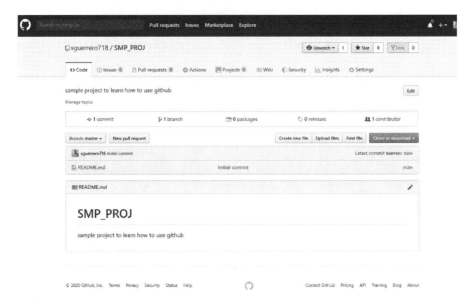

Figure 1-18. *GIT repository sample*

4) Using the earlier Node JS file from Visual Code, set
    that project as the SMP_PROJ repository as shown in
    Figure 1-19. Copy the URL.

Figure 1-19. *Cloning a GIT repo*

From Visual Code, set up the remote repository using the git command
git remote add (Figure 1-20).

```
JS index.js  ×
JS index.js  >  {} http.createServer() callback
   1     // ***************************************
   2     // author: Sergio Guerrero
   3     // hello world node app
   4     // ***************************************
   5
   6     // to run, open the cmd line and type
   7     // node index.js
   8     var http = require("http");
   9
  10  v  http.createServer(function (request, response) {
  11
  12         // response headers
  13         response.writeHead(200, {'Content-Type': 'text/plain'});
  14
  15         // and response message
  16         response.end('Hello World');
  17     }).listen(8081);
  18
  19     // log on the console
  20     console.log('Server running at http://127.0.0.1:8081/');

PROBLEMS   OUTPUT   DEBUG CONSOLE   TERMINAL

Microsoft Windows [Version 10.0.18362.720]
(c) 2019 Microsoft Corporation. All rights reserved.

C:\Users\Sergio\Documents\Tutorials\nodejs>git remote add origin https://github.com/sguerrero718/SMPL_PROJ.git
```

Figure 1-20. *Adding remote repo from VS Code*

After running the git remote add command, verify the remote repository (Figure 1-21).

Figure 1-21. *Verifying git repo version*

Proceed with committing the initial set of files, index.js, and package. json into the GitHub repository that was mapped. Using Visual Code IDE (or another preferred IDE) provides a message describing the changes instead of using git command line at this point. Notice the top-right corner of the next image to see the commit message, and listed are the pending changes to be committed – make sure to ***always*** stage changes prior to committing them; otherwise, Visual Code will prompt for staging them beforehand – and then commit (click the check icon under the Terminal menu) as shown in Figure 1-22.

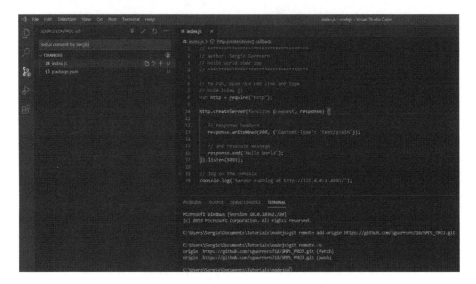

Figure 1-22. *Code commit comment from VS Code*

If the commit is successful, there should be a message, and there should not be any pending changes after that operation (Figure 1-23).

Figure 1-23. *VS Code after commit*

At this point, changes have been staged and committed (they were put in a placeholder). The next step is to push these changes to the remote repository. Select the three dots on the **Source Control: GIT** section and select **Push** or **Push to** in order to select the desired remote repository as shown in Figure 1-24.

Figure 1-24. *VS Code pushing to GIT repo*

An error was received because there must first be a pull followed by a merge and push. The next step is to pull the changes (none initially), then redo the stage, commit, and push. After a few attempts, the committed files appeared in the GitHub repository as displayed in Figure 1-25.

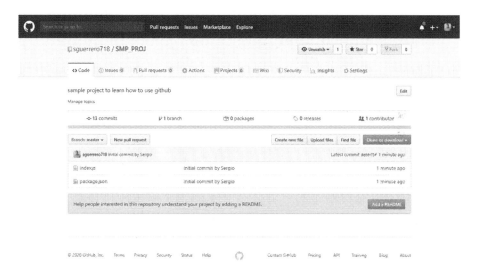

Figure 1-25. *Verifying committed code in GitHub*

When going back to Visual Code, an intentional change was made in the index file by adding a comment. After saving the file, it showed as a pending change. Steps to stage and commit were repeated. Finally, the changes were pushed to the external repository. After a few seconds, those changes were reflected in the commits count (incremented from 13 to 14) as displayed in Figure 1-26.

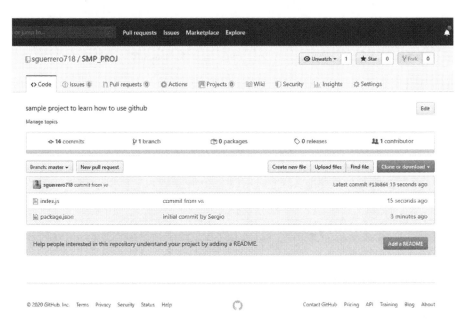

Figure 1-26. *Verifying latest commit*

To verify the file content, see the added comment on line 5 displayed in Figure 1-27.

Figure 1-27. *Verifying code file changes*

Once the SAP Web IDE is mapped to a GitHub repository, the need
to merge and perform all of the additional steps should be unnecessary.
These steps were for illustration purposes to show how different IDEs
can connect to an external code repository like GitHub and how to
make changes to them and then commit those changes by using the git
commands or the IDE plugin interfaces. The SAP Web IDE example will
contain similar screenshots and explanation of the steps on that tool.

mta.yaml

Within cloud applications, there is a file with a file extension yaml (or
yml). YAML stands for "YAML Ain't Markup Language." This is a human-
readable data serialization file standard for all programming languages.
This file specifies how the (containerized) application is built, integrated,
and deployed to a cloud environment. This file has a specific structure, and
it follows strict rules on spacing, naming conventions, syntax, and rules
that can be found on their website (`https://yaml.org/`).

Moreover, within the scope of SAP HANA XSA and from a microservice, this mta (multitarget applications) file explains the module relationship, any services (managed or user provided) that are consumed, and dependencies within a cloud system in order to deploy an application. YAML files contain extremely sensitive syntax and spacing rules; therefore, it is important to understand it before making any modifications to this file.

There are other metadata properties that can be included in this file such as the ones displayed in Table 1-6 and Figure 1-28.

Table 1-6. *mta yaml*

Metadata property	Meaning
ID	Refers to the name of the multitarget application
Version	Refers to the full application (build) version to be deployed. You may do as many builds using the same version; however, when the version is changed, a new mtar file is created
Modules	Refers to the code that makes up the database module, Node JS module, HTML5 module, or Java module
Resources	Something required by MTA but not provided by MTA. This section specifies the use of custom user-provided services (CUPS) or a managed service (service provided by SAP HANA)
Properties	Key-value pairs within the modules used at runtime
Parameters	Reserved variables
Requires	Refers to a module dependency

```
mta.yaml ×
1    ID: wavepress
2    _schema-version: '2.1'
3    version: 0.0.1
4 ▾ modules:
5 ▾  - name: db
6       type: hdb
7       path: db
8 ▾     requires:
9 ▾       - name: hdi_db
10 ▾         properties:
11            TARGET_CONTAINER: '~{hdi-container-name}'
12 ▾       - name: WAVEPRESS_SVC
13          group: SERVICE_REPLACEMENTS
14 ▾         properties:
15            key: ServiceName_1
16            service: ~{wavepress-service-name}
17
18 ▾ resources:
19 ▾  - name: hdi_db
20 ▾     properties:
21         hdi-container-name: '${service-name}'
22       type: com.sap.xs.hdi-container
23 ▾   - name: WAVEPRESS_SVC
24 ▾     parameters:
25         service-name: CROSS_SCHEMA_SVC
26 ▾     properties:
27         wavepress-service-name: ${service-name}
28       type: org.cloudfoundry.existing-service
29
```

Figure 1-28. *mta yaml file*

Conclusion

At the end of this chapter, you should be able to understand the basic SAP HANA 1 environment and be able to understand some of the changes that are coming into SAP HANA XSA, such as the need for an external code repository as GitHub, the introduction to the XSA command-line interface, and also the new artifacts present in the XSA environments, as an introduction to being able to create microservices in HANA XSA.

CHAPTER 2

Security model in XSA

This chapter explains the Cloud Foundry security model using the User Account and Authorization service (UAA) and how SAP HANA XSA implements it. The OAuth protocol and its vital role within XSA are part of this chapter. HDI containers along with cross-schema access on classic objects are explained along with other development artifacts such as synonyms, and grants are addressed as they are vital to the cross-schema development scenarios that are present in any enterprise who is ready to face the implementation of development in an XSA environment.

User Account and Authorization service (UAA)

The User Account and Authorization service is a managed service within the Cloud Foundry architecture. It is used to authenticate requests against a system acting as an OAuth agent, issuing tokens to client applications, and in turn, these tokens are used for authentication and authorization against back-end systems based on some scope linked to the token and user. This (UAA) service helps in the process of logging in users and can also act as a single sign-on authority. In my opinion, this is the most important step in this environment's architecture since without it, it would be impossible to authenticate any user request. Without an authenticated

© Sergio Guerrero 2020
S. Guerrero, *Microservices in SAP HANA XSA*, https://doi.org/10.1007/978-1-4842-6118-7_2

request, an application or API would result in it being inaccessible in a production environment where security is the most important aspect of an entire software system.

In SAP HANA XS advanced, this UAA managed service generates access tokens in the form of JSON Web Tokens (JWTs). A JWT is a compact and self-contained way for securely transmitting information between parties as a JSON object (RFC 7519). The JWTs in SAP HANA XSA contain additional information for authorization and scope for the users.

The common use cases of JWTs are

1) **Authorization** – It is the most common scenario of JWT. Once a user is logged in, each subsequent request will include the token allowing the user to access resources based on its scope. Single sign-on is a feature that has widely used JWTs for many years.

2) **Information exchange** – It is a good way to securely transmit information between parties. JWTs can also be signed using public/private key pairs to validate the authenticity of the request. Additionally, the signature is calculated using the header and payload to verify that the content has not been altered from the original sender to the receiver.

OAuth 2.0

OAuth is the industry-standard authorization protocol used, among other means, by the UAA service for authorization. OAuth 2.0 focuses on client developer simplicity while providing specific authorization flows for web, desktop, and mobile applications and IoT (Internet of Things) devices.

Scope is a mechanism in OAuth 2.0 that limits an application access to a user account. An application can request one or more scopes. Then, the access token issued to the application will be limited to the scopes granted to the user. In layman's terms, scopes define the actions that can be performed within a service or application, for example, being able to perform create, update, delete, and read against a REST API. Before requesting scopes and accessing directly from these authorization servers, developers must (or at minimum try to) think about all the possible ways that can prevent unauthorized access of applications. Developers must follow simple rules that are now enforced by most companies, for example, Google, Facebook, AWS.

When creating new users, it is normal to see that these new accounts by default have

- Minimum permissions (or no permissions at all as is done with new users in AWS)

- Incremental authorization (the process of requesting only what is needed and nothing extra)

- Restricted scopes (only provides access to what is needed to achieve someone's daily tasks and it must be proven)

- Security assessments (in place to minimize the unauthorized requests of too many scopes for certain accounts. Some assessments are costly to immediately eliminate hacking practices)

Through history and through experience, these rules make more sense. Developers usually want to have ALL types of permissions so that they do not repeatedly request access from administrator teams; however, these rules and practices are in place to minimize risk of exposing data to unauthorized users and protect data privacy.

Coming back to the definition of the protocol, the OAuth protocol is used to let a client application know of its authorization on a system, not necessarily who the user is or what they want to do. A perfect example described on the Oauth.net website is a comparison using a hotel customer.

This customer arrived at the hotel and presented his/her identification and payment method (credentials). In turn, someone at the hotel lobby (authorization server) gave the customer a room key (token), and this key is then approved to open a door (some API). This key is not used to open all doors but only the door authorized for the user, only during his/her stay (it will eventually expire, and the same key will not work), as shown in Figure 2-1.

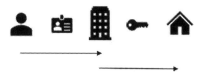

Figure 2-1. *OAuth analogy*

Along with OAuth tokens, there are additional safeguards that developers can include in their applications such as cross-origin resource sharing (CORS) and cross-site request forgery (CSRF).

With CORS, enabling or disabling access to specific resources can be achieved based on the domain. CORS settings can be modified to set specific origins for access or denial of access.

In the application router configuration (in the xs-app.json file in the HTML5 module), developers would need to specify which routes need to be protected. The CSRF setting is called xsrfProtection (default is true). These protections apply to any HTTP request that changes state on the server side (PUT, POST, DELETE).

There are different ways in which JWTs are generated and used. The most common type of JWT OAuth 2.0 access is **bearer tokens**. A bearer token is a long encrypted string that may not have any meaning to clients. Some servers may issue short string hexadecimal tokens, while others may structure tokens as JWTs. We will see this token in Chapter 4.

Custom user-provided services – CUPS

Another type of service within the Cloud Foundry architecture is the
custom user-provided service (CUPS). These CUPS are services that,
by definition, are created by a developer user, not by the framework.
Moreover, these services are created because other services do not exist as
provided by the marketplace. Once created, CUPS will behave similarly to
services created through the marketplace. Its service instances will enable
developers to configure their applications with these, using app binding –
the same app runtime environment variables used by CF to automatically
deliver credentials for marketplace services. In the XSA environment, these
services can be seen from the SAP HANA XSA cockpit or by utilizing the XS
CLI from the Linux terminal.

One of the uses of creating user-provided services in SAP HANA XSA is
to be able to access classic schema tables from database modules.

Let's begin by creating a schema in the SYSTEM database (or a
different schema of your choice) as shown in Figure 2-2.

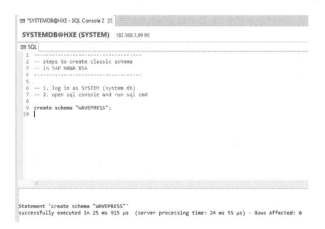

Figure 2-2. *Creating schema in SAP HANA*

Once a schema is created, proceed to create a database table within
this schema as shown in Figure 2-3.

43

```
11  -- 3. create a table under the same schema
12⊖ create column table "WAVEPRESS"."DEVICES" ("ID" INT NOT NULL PRIMARY KEY GENERATED BY DEFAULT as IDENTITY COMMENT 'Identity',
13                                             "DE_UUID" NVARCHAR(20) COMMENT 'Device Identifier',
14                                             "TEMP_F" DECIMAL(10,2) COMMENT 'Temperature Fahrenheit',
15                                             "REC_DT" SECONDDATE DEFAULT CURRENT_TIMESTAMP COMMENT 'Record timestamp');
```

```
Statement 'create column table "WAVEPRESS"."DEVICES" ("ID" INT NOT NULL PRIMARY KEY GENERATED BY DEFAULT as ...'
successfully executed in 111 ms 861 μs  (server processing time: 110 ms 347 μs) - Rows Affected: 0
```

Figure 2-3. *Creating a table in SAP HANA*

Once the table is created, verify it. Run a select statement to see the initial empty table, then add data to this table. For the purpose of this exercise, a single insert statement ran several times. The console log shows the successfully inserted statements as displayed in Figure 2-4.

```
16  -- verify your table (0 records)
17  select * from "WAVEPRESS"."DEVICES";
18
19  -- insert data to your table
20⊖ INSERT INTO "WAVEPRESS"."DEVICES" ("DE_UUID", "TEMP_F", "REC_DT")
21  SELECT '2020-DEA333-3', RAND() * 100, NOW() FROM DUMMY
22
```

```
Statement 'INSERT INTO "WAVEPRESS"."DEVICES" ("DE_UUID", "TEMP_F", "REC_DT") SELECT '2020-DEB349-2', RAND() * ...'
successfully executed in 4 ms 891 μs  (server processing time: 2 ms 416 μs) - Rows Affected: 1

Statement 'INSERT INTO "WAVEPRESS"."DEVICES" ("DE_UUID", "TEMP_F", "REC_DT") SELECT '2020-DEC349-1', RAND() * ...'
successfully executed in 4 ms 687 μs  (server processing time: 2 ms 403 μs) - Rows Affected: 1

Statement 'INSERT INTO "WAVEPRESS"."DEVICES" ("DE_UUID", "TEMP_F", "REC_DT") SELECT '2020-DEC349-2', RAND() * ...'
successfully executed in 5 ms 363 μs  (server processing time: 2 ms 785 μs) - Rows Affected: 1

Statement 'INSERT INTO "WAVEPRESS"."DEVICES" ("DE_UUID", "TEMP_F", "REC_DT") SELECT '2020-DEC349-3', RAND() * ...'
successfully executed in 4 ms 656 μs  (server processing time: 2 ms 456 μs) - Rows Affected: 1
```

Figure 2-4. *Adding sample data*

Verify the table one more time to make sure the records were indeed created successfully as shown in Figure 2-5.

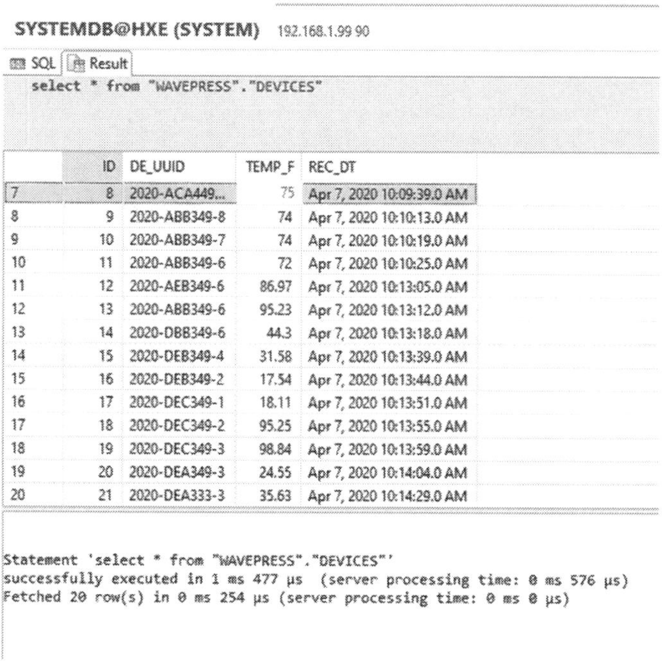

SYSTEMDB@HXE (SYSTEM)   192.168.1.99 90

SQL | Result

select * from "WAVEPRESS"."DEVICES"

	ID	DE_UUID	TEMP_F	REC_DT
7	8	2020-ACA449...	75	Apr 7, 2020 10:09:39.0 AM
8	9	2020-ABB349-8	74	Apr 7, 2020 10:10:13.0 AM
9	10	2020-ABB349-7	74	Apr 7, 2020 10:10:19.0 AM
10	11	2020-ABB349-6	72	Apr 7, 2020 10:10:25.0 AM
11	12	2020-AEB349-6	86.97	Apr 7, 2020 10:13:05.0 AM
12	13	2020-ABB349-6	95.23	Apr 7, 2020 10:13:12.0 AM
13	14	2020-DBB349-6	44.3	Apr 7, 2020 10:13:18.0 AM
14	15	2020-DEB349-4	31.58	Apr 7, 2020 10:13:39.0 AM
15	16	2020-DEB349-2	17.54	Apr 7, 2020 10:13:44.0 AM
16	17	2020-DEC349-1	18.11	Apr 7, 2020 10:13:51.0 AM
17	18	2020-DEC349-2	95.25	Apr 7, 2020 10:13:55.0 AM
18	19	2020-DEC349-3	98.84	Apr 7, 2020 10:13:59.0 AM
19	20	2020-DEA349-3	24.55	Apr 7, 2020 10:14:04.0 AM
20	21	2020-DEA333-3	35.63	Apr 7, 2020 10:14:29.0 AM

Statement 'select * from "WAVEPRESS"."DEVICES"'
successfully executed in 1 ms 477 µs (server processing time: 0 ms 576 µs)
Fetched 20 row(s) in 0 ms 254 µs (server processing time: 0 ms 0 µs)

Figure 2-5. *Verifying inserted data*

At this point, there is enough in the table to simulate a classic SQL schema. Next, analyze the XSA project that needs to be created. Start by creating a new project of type MTA (multitarget application).

Remember from the Cloud Foundry hierarchy structure, applications belong to a single space. During this initial step, select the desired XSA space to bind the application to during the project creation. Notice also the application version (initial version 0.0.1). The application version is helpful during deployments in order to deploy new features or apply bug fixes. For now, keep this application version as 0.0.1.

Click Finish and Confirm after Figure 2-6.

Figure 2-6. *Creating a new mta project*

Once the project has been created, it will display under the Workspace in the SAP Web IDE with the default structure that includes the mta.yaml file similar to Figure 2-7.

Figure 2-7. *Displaying a new project in the Workspace*

Let's get to work. Start by creating a database module called db so that it can hold database models, stored procedures, and synonyms, among other artifacts, as shown in Figure 2-8.

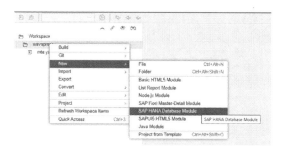

Figure 2-8. *Creating a database module*

The first step is to provide a module name. For simplicity, name it **db** as displayed in Figure 2-9.

Figure 2-9. *Naming the database module*

Then the **db** module wizard will ask for an optional namespace, a schema name, and the preferred SAP HANA database. At this time, use version HANA 2 SPS 04 to take advantage of the latest features and development in the environment as shown in Figure 2-10.

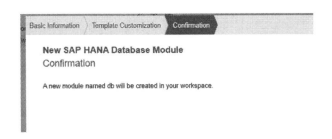

Figure 2-10. *Selecting the database version*

Click next and then finish as shown in Figure 2-11. If everything is provided correctly, the confirmation message should appear, and after closing it, the newly created module will display inside the wavepress project.

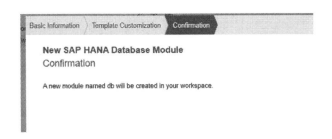

Figure 2-11. *Confirmation of db module creation*

Since the checkbox to do an initial build of the database module was ticked, it may take a few seconds to create the module and run the first **db** module build. This is done to ensure the **db** module and the HDI (HANA Deployment Infrastructure) container were created correctly. Before adding any more development into the project, all should be inline and with no issues in the initial installation and project creation. The SAP Web IDE console will show the build log, and eventually a successful build message should appear. With the conclusion of this process, a number of things happened in the background. The **db** module was built, the mta. yaml file was updated with the HDI container, the HDI namespace was updated, some container users were created in the database, and, finally, the completion log message was received. At this point, much of what happened may not be visible from the Web IDE. It is OK; these items can be verified later. For now, assume it was all done as expected by the Web IDE as shown in Figure 2-12.

Figure 2-12. *Web IDE console build message*

If the **db** module is expanded inside the project, there is an **src** folder and a package.json file that were created. The **src** folder will be used to house development artifacts (such as models, stored procedures, synonyms), while the package.json file is used every time a **db** module build is triggered. There is metadata inside of this file that is used during the build of the project including dependencies and build information as shown in Figure 2-13.

Figure 2-13. *SAP Web IDE showing db module and package.json inside the db module*

So far, the only work that has been completed is with the database module. Wouldn't a developer expect to see a database screen at some point? Well yes! Notice the left navigation panel of the SAP Web IDE; so far everything has been done in the Workspace section which is represented with an xml tag icon.

The **Database Explorer** section is the very next icon (represented as four cubes) down from the workspace icon. Click it and see what happens. In other database systems, the first time a database is opened, it prompts for connection string settings and credentials. It is no different from the SAP HANA Database Explorer. Since there has not been any database connection configured at this point and the SAP Web IDE knows that the current project is working with an HDI (HANA Deployment Infrastructure) container (created during the db initial module build and there has not been any schema specified to be used during the database module creation), it will ask if it should add one now. The option at this point is to map to the default HDI container, or, if any schema credentials are known, then that connection may also be established as shown in Figures 2-14 and 2-15.

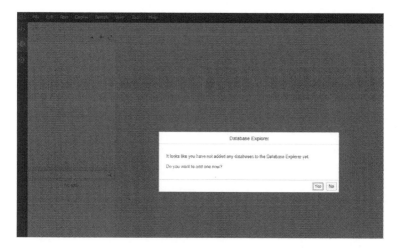

Figure 2-14. *Opening the Database Explorer for the first time*

Start with the default HDI container. Notice the HDI container name is made up of the username, some unique identifier, the project name, hdi_ db, and the space name in parenthesis. If more than one HDI container connection is created, a similar pattern will be followed. Proceed with the default setting at this point. This HDI container will have any CDS models (hdbcds files) and synonyms we create from our database module. On the other hand, a classic schema connection will contain any database objects from the classic world.

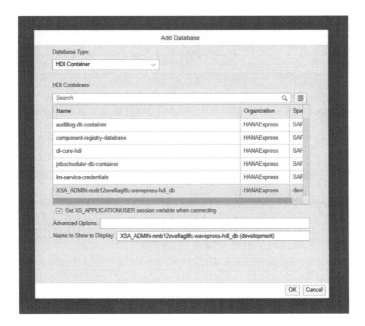

Figure 2-15. *Adding a database connection (showing the HDI container)*

If the HDI container connection is established correctly, it should look like this. Select the *Tables* section and notice there are no tables created initially by the HDI container. Later, after tables have been created, they will be displayed in this section. If the table list is too long, it can be refined by typing a few characters in the Search box at the bottom of Figure 2-16.

Figure 2-16. Showing the HDI container from the database connection

Now, the **db** module has an HDI container and has been built correctly. In a demo scenario, these may be enough steps for someone to create models, see tables, and query data. In a real scenario, that is not the case. In a real scenario, a developer will need to consume and integrate data from logically separated database tables or schemas. A developer may need to create mixed information models and serve that data from a calculation view or query them from stored procedures or table functions. How would a developer achieve such a scenario given that only an HDI container and the database module have been created so far? Keep in mind that the SAP HANA XSA environment is following Cloud Foundry principles and that developers must also need to understand security between containers and the container isolation concept; accessing separate schemas is not straightforward as it once was in SAP HANA 1.

In SAP HANA XS advanced architecture, custom user-provided service (CUPS) needs to be created and access granted to a schema to our CUPS service. Before SAP HANA XSA SPS 03, this step had to be done from the XS command line using the *xs cups* command *xs cups* {CUPS_NAME} – "host", "port", "user", "password", "driver", "tags", "schema" (very tedious process if a developer was not familiar with the command syntax and also if the developer had no prior experience with cli tools)

There is a new feature of the **db** module called **SAP HANA Service Connection** – illustrated in Figure 2-17 using SAP HANA XSA SPS 04. To reach this feature, right-click the **db** module, select New, and then SAP HANA Service Connection.

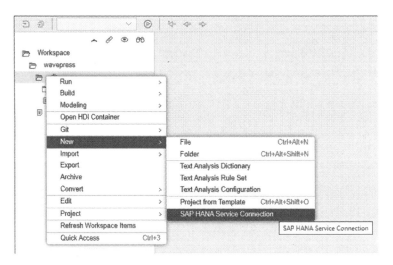

Figure 2-17. *Adding SAP HANA Service Connection (custom service)*

When selecting this feature (Figure 2-17), there will be a wizard that prompts for CUPS settings such as a name, connection details to the database, and credentials for authentication as shown in Figure 2-18. Be cautious when creating this CUPS service as any settings that are erroneous due to incorrect authentication will not be shown to the user until the db module is rebuilt. Once this step is completed, settings will be updated in the mta.yaml file.

Figure 2-18. *Custom service properties*

Open the mta yaml file to see the new section added. Notice the new CUPS service appears inside the Resources section. If the credentials provided are incorrect, an error message for the CUPS service will be displayed in the console following the building of the **db** module. Remove the newly added Resource (CUPS service) and the newly required service within the **db** module. Once this issue is resolved, go back and add a new SAP HANA Service Connection until getting a successful CUPS connection. A successful connection is determined when an error message is no longer shown. This step may take multiple attempts before being successful. Here are some tips to help with the successful creation of a SAP HANA Service Connection:

1) Before adding the CUPS, ensure the **db** module always builds successful – it will save a lot of debugging time later. Do not try to debug many issues at once; divide and conquer!

2) When adding the CUPS service (from the **db** module > New > SAP HANA Service Connection), immediately open the mta yaml file to see its structure. Notice the Resources section and the new item created (of type org.cloudfoundry.existing-service).

3) Knowing that the **db** module will need the newly created resource as a required module – this is the only way the **db** module can correctly connect to the classic schema – carefully make sure the required section uses the name of the CUPS resources defined.

4) Having patience with the process will make development more successful.

All previous steps will result in Figure 2-19.

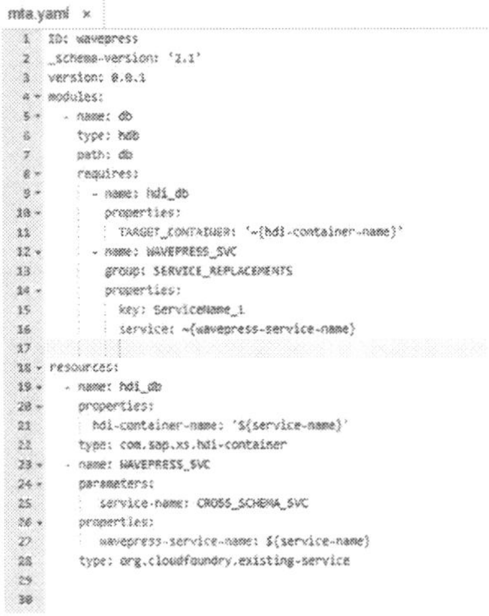

Figure 2-19. *mta file with the db module and custom service details*

Cross-schema access using synonyms

Synonyms allow objects to call something by a different name. Within SAP HANA XSA, the system makes use of these synonyms in order to be able to read from HANA classic schemas (non-HDI containers). A HANA synonym (hdbsynonym file) will need to be created in order to use the DUMMY table as it is coming from the SYSTEM's schema. Synonyms will work if they receive grant access from the source data to the consumer of the synonym. Synonyms by themselves may appear to be unnecessary; however, their importance will be showcased in the following.

After adding the CUPS service and rebuilding the **db** module again, schema privileges will need to be assigned (granted) to this CUPS service. By the schema owner granting access, this account gives permissions to query the schema WAVEPRESS via an object called hdbgrants as shown in Figure 2-20.

Figure 2-20. *hdbgrants*

The hdbgrants file is a JSON formatted file. It specifies an **object_owner** by its name "ServiceName_1" which comes from the mta.yaml file. It further represents the WAVEPRESS_SVC CUPS service that was created as a SAP HANA Service Connection. Along with the binding of the service name to the CUPS, the object owner also assigns schema privileges (SELECT, SELECT METADATA, CREATE, UPDATE, DELETE).

The following section within the hdbgrants file acknowledges the **application_user** that will receive the schema_privileges referencing a database schema and the privileges that the grantee will be able to apply during development.

After adding this hdbgrants file and saving it, the **db** module needs to be rebuilt correctly without introducing any syntax errors before proceeding.

Select the **db** module. Right-click and select Build. If it successfully builds, then proceed; otherwise, go back and adjust the CUPS service, the mta file, or the hdbgrants file. This may take a few iterations to complete correctly due to the various integration points and sensitive syntax.

Carefully read the SAP Web IDE console for error message details. Some errors encountered may be due to incorrect credentials during the CUPS creation. If an error is encountered, repeat the steps provided earlier. If the CUPS service is renamed and it is unknown whether the old service still exists (not visible from the SAP Web IDE), navigate to the XSA cockpit to see the service instances that exist in the current system.

If the service name is changed while recreating these CUPS, orphan services will exist in the cockpit. If this occurs, proceed to the cockpit, navigate to the service instances, and perform due diligence in cleaning up those services.

The next step is to create a synonym to select data from the cross-schema table called WAVEPRESS.DEVICES as displayed in Figure 2-21.

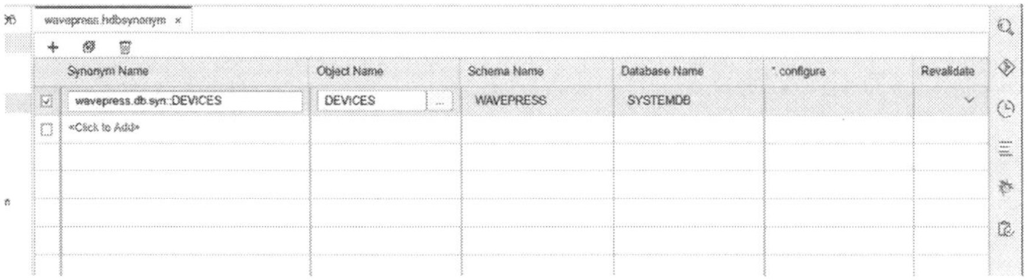

Figure 2-21. *hdbsynonym files*

Repeat the previous step and create a synonym for the DUMMY table to be added within the same hdbsynonym file (as well as others that may be needed later) as shown in Figure 2-22. Every time there is a modification to a db artifact, and development is ready to be validated, build the db module to ensure these artifacts are updating correctly. If the synonyms were created correctly, switch to the SQL console and test the newly created synonyms.

Synonym Name	Object Name		Schema Name	Database Name	*configure	Revalidate
☑	wavepress.db.syn::DEVICES	DEVICES	...	WAVEPRESS	SYSTEMDB		⌄
☐	wavepress.db.syn::DUMMY	DUMMY	...	SYS			⌄
☐	<Click to Add>						

Figure 2-22. *Multiple tables within the same synonym file*

At this point, a CUPS service, a grants file, and a synonym have been created. It is now appropriate to create a calculation view to expose the database table data.

Return to the **db** module and create a new folder called **models**. Inside this folder, a new information model will be created. Right-click the folder, select New, and then Calculation View, and provide a meaningful name when prompted following Figure 2-23.

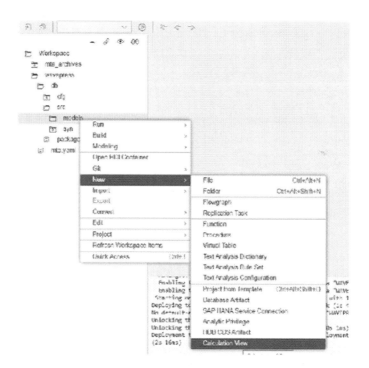

Figure 2-23. *Creating calculator view*

Since the DEVICES table will need to be exposed, the model name will be CV_DEVICES. By default, this view will be created as a CUBE view (which requires an aggregated node before the semantic layer). Leave it as is and proceed after Figure 2-24.

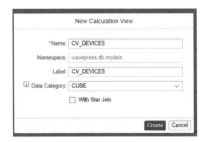

Figure 2-24. *Calculation view initial properties*

When performing database modeling, developers and consumers of the model do not need to know or should not care where the data is coming from as long as they can query the table or view. At a later time, additional nodes will be included from the graphical editor after the view is created.

Add a projection node from the middle section of the screen (calculation view node types). When the projection is added, there are some icons on the node, select the + (plus sign) to find the desired data source – WAVEPRESS.DEVICES table. Voila! Select the table and add it to the projection as shown in Figure 2-25.

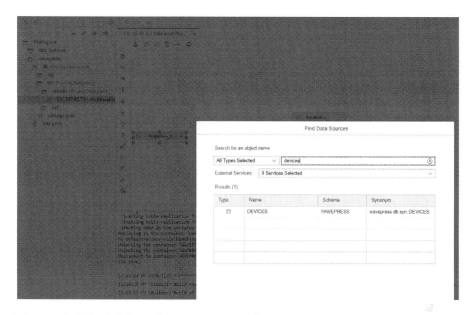

Figure 2-25. *Adding data source to the projection in calculation view*

As more nodes are added into the graphical view, the model will eventually need to be realigned for better visibility of itself. Under the name of the calculation view, some icons are displayed to help with the layout, debugging, and performance analysis. Select the auto layout icon to make the data model visually more attractive. It should now look like Figure 2-26.

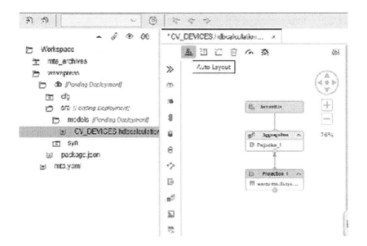

Figure 2-26. *Auto layout function*

After adding a data source in the Projection_1 node, map the columns upward. Drag columns from the Data Sources box into the Output Columns, or if all the columns need to be projected, select the **Add To Output** icon from within the Mapping section as shown in Figure 2-27.

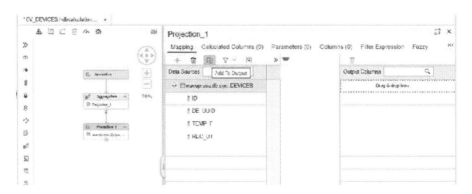

Figure 2-27. *Mapping out columns within the projection*

Repeat the same **Add To Output** if requiring the same columns moving upward as modeling progresses upward during the data model build. The mapped-out columns will show as a line from source to target. Should calculated columns be required, these may also be included by going to the Calculated Columns tab and adding them as needed as shown in Figure 2-28.

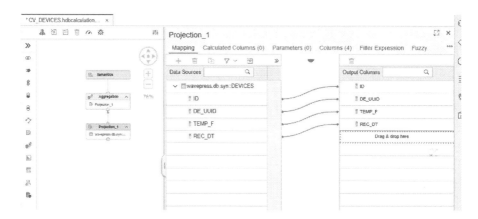

Figure 2-28. *Column mappings*

During the view creation, the CUBE view is the default behavior that was set from the initial creation wizard. It is now appropriate to change the Data Category to a Dimension view (not aggregated). By selecting the Aggregation node before the semantic layer and clicking it (Figure 2-29), select Switch to Projection so adding a measure to the output is not required. Additional features, such as Data Previewing, are also displayed on the context menu. These features appear on a node-by-node basis. The Data Preview feature is very helpful during data model debugging – described in Chapter 3.

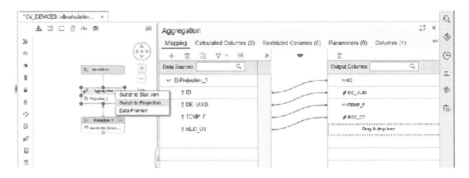

Figure 2-29. *Column mapping in the aggregation node*

As development moves from the bottom Projection_1 node up to the semantic layer, the Columns and Labels (comment added during table creation) to be outputted are seen as well as their data type expected (Measure vs. Attribute) as shown in Figure 2-30.

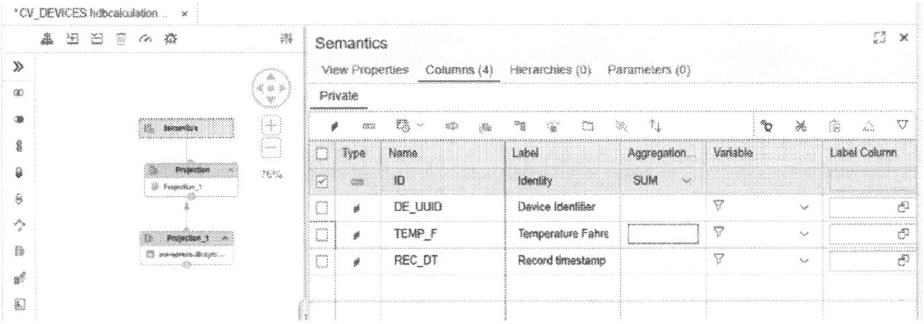

Figure 2-30. *Semantic layer to set data aggregation and column type*

Another helpful feature in the calculation view is the ability to rename the nodes. Using meaningful names during the development is helpful to understand what is done in each node. The bottom node was renamed from Projection_1 to p_devices as it represents a projection of the devices table. It should be noted that in the semantic layer view properties, the Data Category property was set as DIMENSION (from default CUBE view setting) as shown in Figure 2-31.

Figure 2-31. *Semantic layer view properties*

When the changes to the view are satisfactory, right-click the db module again and build it. If the db module builds successfully, then ensure that it can read data from the view.

Right-click the view, and select Data Preview – this is an easy way to generate a sample view of the calculation view data without having to type the entire query of "select * from view" in the SQL console as shown in Figure 2-32.

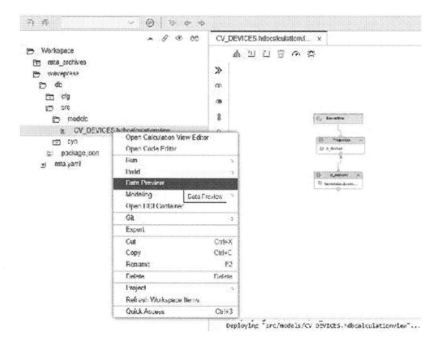

Figure 2-32. *Running a Data Preview of the view*

Select the "Raw Data" tab as shown in Figure 2-33.

	ID		DE_UUID		TEMP_F		REC_DT
1	2		2020-ABC999-1		73.00		2020-04-07 10:06:00
2	3		2020-ABC999-2		71.00		2020-04-07 10:06:00
3	4		2020-ABC999-3		73.00		2020-04-07 10:06:00
4	5		2020-ABC999-5		73.00		2020-04-07 10:09:00
5	6		2020-ABC949-3		71.00		2020-04-07 10:09:10
6	7		2020-ACC949-3		78.00		2020-04-07 10:09:25
7	8		2020-ACA449-3		75.00		2020-04-07 10:09:39
8	9		2020-ABB349-8		74.00		2020-04-07 10:10:13
9	10		2020-ABB349-7		74.00		2020-04-07 10:10:19
10	11		2020-ABB349-6		72.00		2020-04-07 10:10:25
11	12		2020-AEB349-6		86.97		2020-04-07 10:13:05

Figure 2-33. *Displaying the Raw Data window*

After creating the data model, create a stored procedure within the **db** module. Scenarios to simulate create, update, and delete operations will be shown – first from the SQL console and later from the **api** module. Create a folder named procs to hold stored procedures. Within this folder, create a stored procedure and name it SP_DEVICE_UPSERT as displayed in Figure 2-34. The purpose of the procedure is to create a new record if the record does not exist; otherwise, the record should be updated based on the primary key.

Figure 2-34. *Showing the new stored procedure created*

This stored procedure is coded to accept an input parameter as a table type structure, and the output will be the ID of the last inserted or updated record (known as identity). The code shown is very basic for illustration purposes, and it does not handle nonexisting IDs or any error messages. Assume the execution will be done via the happy path. Make sure that the stored procedure does not contain READ SQL DATA so that it can perform the create or update operation against the database table – commented out on line 7 in Figure 2-35.

```
SQL Console 1.sql  ×    SP_DEVICE_UPSERT.hdbpro...  ×
  1 ▾  PROCEDURE "wavepress.db.procs::SP_DEVICE_UPSERT"(
  2         IN IN_DEVICE TABLE("ID" INT, "DE_UUID" NVARCHAR(20), "TEMP_F" DECIMAL(10, 2), "REC_DT" SECONDDATE)
  3       , OUT OUT_ID INT
  4       )
  5     LANGUAGE SQLSCRIPT
  6     SQL SECURITY INVOKER
  7     --READS SQL DATA
  8     AS
  9  BEGIN
 10
 11 ▾    vDevice = SELECT TOP 1 in_d.*
 12              FROM "wavepress.db.syn::DEVICES" d
 13                INNER JOIN :IN_DEVICE in_d ON d."ID" = in_d."ID";
 14
 15 ▾    IF( IS_EMPTY(:vDevice) ) THEN
 16 ▾        INSERT INTO "wavepress.db.syn::DEVICES" ("DE_UUID", "TEMP_F", "REC_DT")
 17          SELECT "DE_UUID", "TEMP_F", NOW() FROM :IN_DEVICE;
 18
 19          SELECT CURRENT_IDENTITY_VALUE() INTO OUT_ID FROM "wavepress.db.syn::DUMMY";
 20        ELSE
⊠21 ▾        UPDATE dd
 22          SET "TEMP_F" = in_d."TEMP_F", "REC_DT" = NOW()
⊠23          FROM "wavepress.db.syn::DEVICES" dd
 24              INNER JOIN :vDevice in_d ON dd."ID" = in_d."ID";
 25
 26          SELECT TOP 1 "ID" INTO OUT_ID FROM :vDevice;
 27
 28      END IF;
 29
 30  END
```

Figure 2-35. *Stored procedure sample code*

Since more privileges will be used (INSERT, CREATE, DELETE), the hdbgrants file needs to be updated to include the additional privileges to avoid errors during the development and build of the db module as shown in Figure 2-36.

```
  SP_DEVICE_UPSERT.hdbpro...   ×   wavepress.hdbgrants  ×   SQL Console 1.sql  ×

 1   {
 2     "ServiceName_1": {
 3       "object_owner" : {
 4         "schema_privileges":[
 5           {
 6             "reference":"WAVEPRESS",
 7             "privileges_with_grant_option":["SELECT", "SELECT METADATA","UPDATE","INSERT","DELETE"]
 8           }
 9         ]
10       },
11       "application_user" : {
12         "schema_privileges":[
13           {
14             "reference":"WAVEPRESS",
15             "privileges_with_grant_option":["SELECT", "SELECT METADATA","UPDATE","INSERT","DELETE"]
16           }
17         ]
18       }
19     }
20   }
```

Figure 2-36. hdbgrants file showing the schema privileges

Having updated these artifacts, now rebuild the **db** module. If it builds successfully, the system will update the hdbgrants configuration behind the scenes, and the stored procedure will be created in the HDI container.

Validate this assumption from the Database Explorer window to see the stored procedure showing under the HDI container connection, under the Procedures section as shown in Figure 2-37.

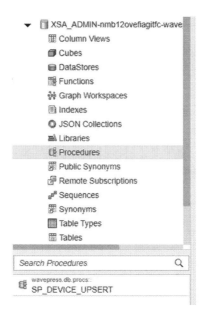

Figure 2-37. *Stored procedure displayed from the Database Explorer*

Once the stored procedure has been created, validate it from the SQL console. In order to unit test this database object, the following needs to be completed:

1) Create a local temporary table that has the similar structure as the input parameter of table type.

2) Add at least one record into this temporary table from the SQL console as shown in Figure 2-38.

```
 2
 3    -- create structure to mimic the input param
 4    create local temporary table #a ("ID" INT, "DE_UUID" NVARCHAR(20), "TEMP_F" DECIMAL(10, 2), "REC_DT" SECONDDATE);
 5
 6 ▾  insert into #a(ID, DE_UUID, TEMP_F, REC_DT)
 7    values(null, '20200413-TEST-2', 54, NOW());
 8
 9    -- validate new record
10    select * from #a;
11
12
13    -- execute sp
14    call "wavepress.db.procs::SP_DEVICE_UPSERT" (#a, ?);
15
```

Messages ×

```
Statement 'create local temporary table #a ("ID" INT, "DE_UUID" NVARCHAR(20), "TEMP_F" DECIMAL(10, 2), ...'
executed in 4 ms.
Statement 'insert into #a(ID, DE_UUID, TEMP_F, REC_DT) values(null, '20200413-TEST-2', 54, NOW())'
executed in 1 ms - Rows Affected: 1
```

Figure 2-38. *Creating some data in the temporary table*

3)   Validate the newly added record (Figure 2-39).

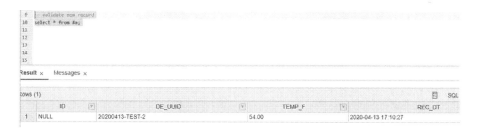

Figure 2-39. *Showing the data from the local temporary table*

4)   Execute the call statement as shown in Figure 2-40.

Figure 2-40. *Calling the stored procedure*

71

5) Validate the table to ensure the latest created record in the table was indeed the one returned as shown in Figure 2-41.

```
16 ▾ select * from "WAVEPRESS_HDI_DB_1"."wavepress.db.syn::DEVICES" order by ID desc
```

Result ×   Messages ×

Rows (22)

	ID		DE_UUID		TEMP_F		REC_D
1	23		20200413-TEST-2		54.00		2020-04-13 17:11:55
2	22		20200413-TEST-1		56.00		2020-04-13 17:03:57
3	21		2020-DEA333-3		35.63		2020-04-07 10:14:29
4	20		2020-DEA349-3		24.55		2020-04-07 10:14:04
5	19		2020-DEC349-3		98.84		2020-04-07 10:13:59

Figure 2-41. *Validating the data that was ran through the stored procedure*

In order to run an update through the stored procedure, perform the next set of steps (Figure 2-42):

1) Remove the existing record from the local temporary table.

2) Add an existing record from the table into the temporary table.

3) In the exercise, it is shown as record ID = 20, changing its TEMP_F value to 26.

4) The stored procedure will also update the REC_DT to current date.

```
6 ▾  insert into #a(ID, DE_UUID, TEMP_F, REC_DT)
7    select ID, DE_UUID, 26, NOW() FROM "WAVEPRESS_HDI_DB_1"."wavepress.db.syn::DEVICES" where ID = 20;
8
```

Figure 2-42. *Populating the temporary table with an existing record to simulate an update*

5) Run the call statement again to validate that the stored procedure updated the desired record (Figure 2-43).

The first output (displayed as the Result1 tab) is from running the call statement, and the OUT_ID column is mapped to the placeholder **?** from the call statement.

Result1 x	Result2 x	Messages x

Rows (1)

		OUT_ID
1	20	

Figure 2-43. *The output of the stored procedure returns the ID of the record that was updated*

The second output window (shown as the Result2 tab in Figure 2-44) displays the select statement that was run. It validates the record that was updated, and it is also displayed on the console in descending order as shown in Figure 2-44. Validate that the TEMP_F value was updated correctly and that the REC_DT shows the latest update first.

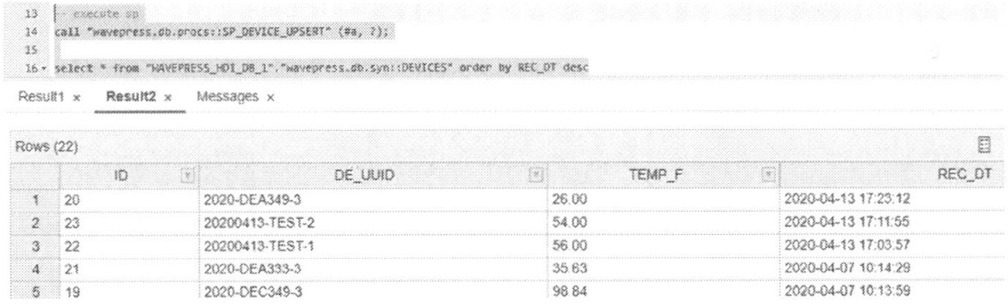

```
13    execute sp
14    call "wavepress.db.procs::SP_DEVICE_UPSERT" (#a, ?);
15
16 ·  select * from "WAVEPRESS_HDI_DB_1"."wavepress.db.syn::DEVICES" order by REC_DT desc
```

Result1 x	**Result2** x	Messages x

Rows (22)

	ID	DE_UUID	TEMP_F	REC_DT
1	20	2020-DEA349-3	26.00	2020-04-13 17:23:12
2	23	20200413-TEST-2	54.00	2020-04-13 17:11:55
3	22	20200413-TEST-1	56 00	2020-04-13 17:03:57
4	21	2020-DEA333-3	35.63	2020-04-07 10:14:29
5	19	2020-DEC349-3	98 84	2020-04-07 10:13:59

Figure 2-44. *Validating the table after the update*

This same stored procedure will be consumed from the Node JS module and demonstrated in a later chapter of the exercise.

The last scenario features how to delete a record. The procedure code to run a delete statement looks like Figure 2-45. *Make sure the code does not contain the line READS SQL DATA so that it can perform the delete operation against the database table.*

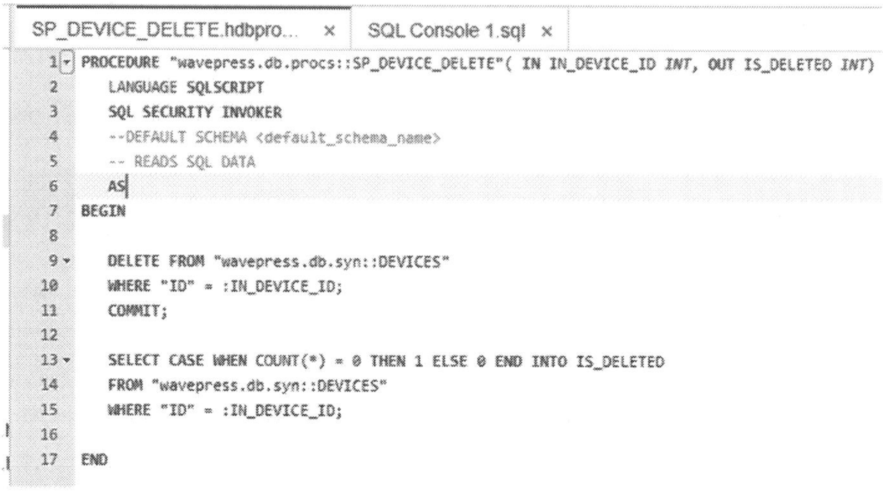

```
SP_DEVICE_DELETE.hdbpro...  ×    SQL Console 1.sql  ×

1   PROCEDURE "wavepress.db.procs::SP_DEVICE_DELETE"( IN IN_DEVICE_ID INT, OUT IS_DELETED INT)
2      LANGUAGE SQLSCRIPT
3      SQL SECURITY INVOKER
4      --DEFAULT SCHEMA <default_schema_name>
5      -- READS SQL DATA
6      AS
7   BEGIN
8
9      DELETE FROM "wavepress.db.syn::DEVICES"
10     WHERE "ID" = :IN_DEVICE_ID;
11     COMMIT;
12
13     SELECT CASE WHEN COUNT(*) = 0 THEN 1 ELSE 0 END INTO IS_DELETED
14     FROM "wavepress.db.syn::DEVICES"
15     WHERE "ID" = :IN_DEVICE_ID;
16
17  END
```

Figure 2-45. *Stored procedure showing a delete operation*

Once the db module is rebuilt successfully, the new stored procedure will appear in the Database Explorer within the Procedures section as shown in Figure 2-46.

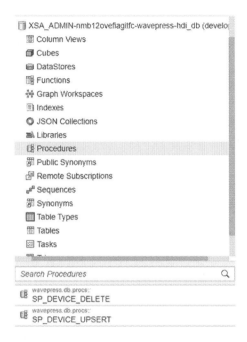

Figure 2-46. *The stored procedure to delete a record shown from the Database Explorer*

Before executing the SP_DEVICE_DELETE statement, run a select statement to display the current records in the table like Figure 2-47.

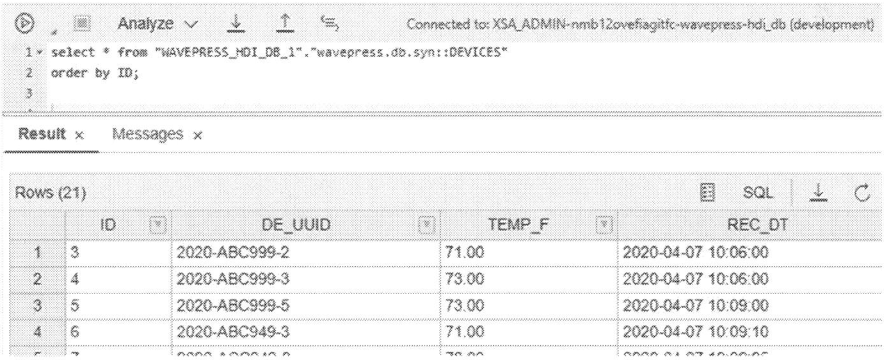

	ID		DE_UUID		TEMP_F		REC_DT
1	3		2020-ABC999-2		71.00		2020-04-07 10:06:00
2	4		2020-ABC999-3		73.00		2020-04-07 10:06:00
3	5		2020-ABC999-5		73.00		2020-04-07 10:09:00
4	6		2020-ABC949-3		71.00		2020-04-07 10:09:10

Figure 2-47. *Table data before running the stored procedure to delete a record*

Next, execute the stored procedure from the SQL console to unit test it, running the call stored procedure statement and passing an ID from the available records (this exercise assumes that valid IDs are always passed to the stored procedure) as shown in Figure 2-48.

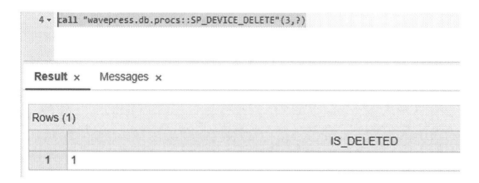

Figure 2-48. *Output after running the stored procedure to delete a record*

Rerun the select statement to validate that the record was successfully deleted as shown in Figure 2-49.

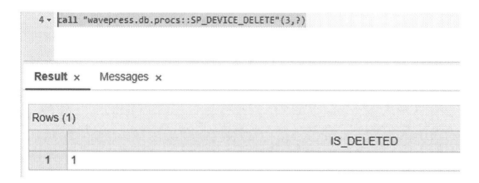

Figure 2-49. *Validating the data from the table after deleting a record*

Keep the stored procedure as is for the time being. It will be used in a later chapter when called from the api module.

Conclusion

At the conclusion of this chapter, you should have some understanding of the CF security layer and how it is implemented in HANA XSA. The OAuth protocol plays an important role as a guard in the development of XSA microservices as the XSA environment follows the same architecture and methodology of requesting and validating tokens. There are some artifacts introduced such as synonyms, user-provided services, and grants that facilitate the access of data on a cross-schema environment.

CHAPTER 3

Tools for development on HANA XSA

This chapter will dive deeper into understanding the tools related to REST API development using NodeJS inside SAP HANA XSA. The tools that will be discussed in this chapter include the SAP Web IDE, the Database Explorer, the XSA cockpit, and GitHub.

SAP Web IDE and connecting to GIT

First, these tools allow developers to develop enterprise software applications. Moreover, they allow debugging the code. Development and debugging go hand in hand. Some developers consider debugging negatively; instead, it should be considered a fundamental step in any development scenario. There is not a single developer in the world that can develop without debugging, even during someone's initial program "hello world." As is necessary with all programming developer tools, the explained below will help developers understand the programming language execution, how the next line runs, how it displays on the console and the browser, or simply how a programming language can complete a task.

Development in SAP HANA XSA is no different. In the initial chapters, it was shown how to use the XS command-line interface

© Sergio Guerrero 2020
S. Guerrero, *Microservices in SAP HANA XSA*, https://doi.org/10.1007/978-1-4842-6118-7_3

which is crucial in the development of SAP HANA XSA. Whether your role is as an administrator, a solution architect, a developer, or a quality assurance engineer, at some point, it will be important to understand the tools involved in the software development cycle. In Chapter 2, a few screenshots from the SAP Web IDE were shown, and now, the book will expand within this section.

It is important to begin by understanding what the SAP Web IDE is. It is a tool developed and provided by SAP via a web browser as their integrated development environment. SAP describes the SAP Web IDE as "A powerful, extensible, web-based tool that simplifies development of end-to-end Fiori applications and the full stack application lifecycle." By default, the SAP HANA Web IDE can be reached from the following URL, http(s)://host:53075 (default port), or from a different port if it was installed in this manner. Consult with the system administrator at your company if the exact location is unknown.

The SAP Web IDE has multiple versions; however, their interface looks similar. Familiarize yourself with the different versions of the SAP Web IDE and their licensing terms to ensure the appropriate version and license are used prior to starting any development. The following are different versions of the SAP Web IDE:

1) The SAP Web IDE on SAP Cloud Platform (Figure 3-1)

2) The SAP Web IDE for SAP HANA (on-premise and HANA Express Edition)

3) The personal SAP Web IDE

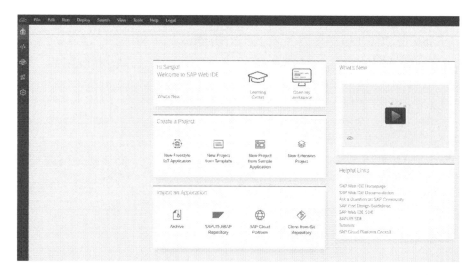

Figure 3-1. *SAP Web IDE*

What are the differences in each environment?

The **SAP Cloud Platform Web IDE** provides templates to create Fiori applications and extensions and to connect to ABAP back-end repositories, and it can also connect to git repositories. When creating an application on the SAP Cloud Web IDE, one of three environment types can be chosen.

The first two types of environments, **ABAP** and **Neo**, have the same types of Fiori (SAPUI5) templates, master detail, list report, worklist, and SAPUI5 application. These two environments are simple enough to start creating user interfaces (UI) with the templates provided. Notice that when creating a project based on one of these templates, the SAPUI5 version needs to be selected. SAPUI5 is a SAP library of modern HTML5 and CSS3 controls that uses JavaScript to create Fiori applications. If deciding to create a SAPUI5 application, check the currently supported versions on the SAPUI5 SDK (Software Development Kit) website.

Other differences between these two environments include the NEO environment (Figure 3-3) being a SAP proprietary environment, and in it, developers can create complex Java, XS JavaScript (XSJS), and HTML5 applications. It is the simplest type of environment to develop in the SAP Cloud. The NEO environment does not support Node JS development.

The ABAP environment in Figure 3-2, as the name refers to, is to create and extend ABAP applications from the SAP Cloud Platform. This environment has licensing costs that need to be investigated prior to using.

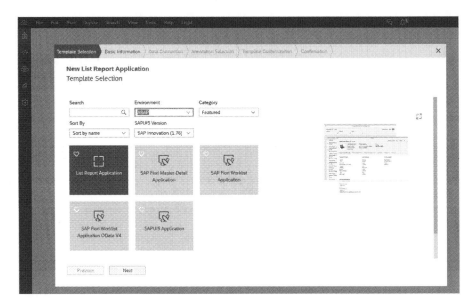

Figure 3-2. *SAP Web IDE from SAP Cloud Platform using the ABAP environment*

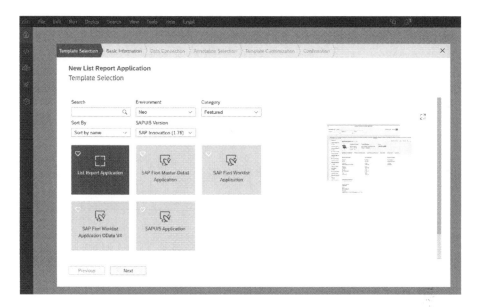

Figure 3-3. *SAP Web IDE Template Selection using the Neo*
environment

The third type of environment is known as the **Cloud Foundry** (CF)
environment as shown in Figure 3-4. Due to the nature of and capabilities
of CF – refer to Chapter 1 for information on CF – there are many other
types of templates when working with the CF environment.

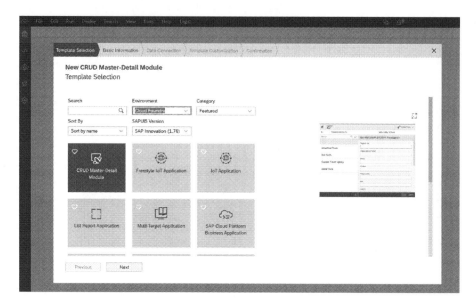

Figure 3-4. *SAP Web IDE Template Selection using the Cloud Foundry environment*

In addition to creating Fiori applications, there are templates for Internet of Things (IoT) applications, templates to create OData services and Cloud Platform Business Applications (CBA). Once an environment and project types are selected, the SAP Cloud Platform Web IDE will appear as shown in Figure 3-5. It is very similar to the on-premise SAP Web IDE version shown in Chapter 2.

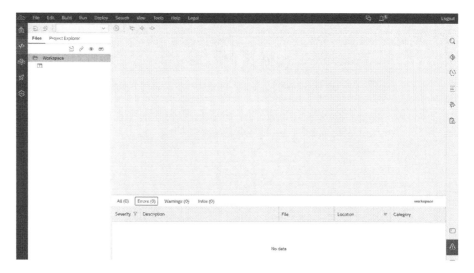

Figure 3-5. *SAP Web IDE displaying the Workspace*

The menu items across the top of the screen are

1) **File** – To create a new file or import a file/project

2) **Edit** – To undo/redo an action and to add comments (line or block)

3) **Build** – To build modules (shown in the previous chapter and shown in the next chapter)

4) **Run** – To run an application or run it as a unit test

5) **Deploy** – To deploy to an ABAP repository or the SAP Cloud or register a Fiori Launchpad

6) **Search** – To make searches on files such as the advanced search on the right navigation icon (magnifying glass)

7) **View** – Allows a developer to view different panes (git, history, debug) or the console, error, and warning windows located on the bottom right hand of the screen

8) **Tools** – Allows switching from the workspace to the Database Explorer, the SAP cloud cockpit, storyboard, preferences, or other features of the SAP Cloud Platform

9) **Help** – Provides links to documentation, the SDK, shortcuts, tips and tricks, About section, and feedback

10) **Legal** – Provides information about terms of use and privacy statements

Notice the icons on the left navigation bar:

1) **Workspace** – It is the workspace used during development, and it is represented by the xml tags icon.

2) **Database Explorer** – To connect to a database or HDI container.

3) **Storyboard** – This feature does not exist on the on-premise version of the SAP Web IDE. This feature is for creating the visual representation of an application and setting up service endpoints used in the application and navigation without writing any code.

4) **Preferences** – This feature is used to set up Web IDE settings such as code editor rules, data previews, git settings, SQL console settings, and workspace settings

The icons on the right navigation bar are

1) **Magnifying glass** – Advanced repository search, used to find content in files within folders, projects, or the workspace.

2) **Git pane** – Used to connect and interact with the
   external code repository (perform actions against the
   code repository such as pull, push, commit, rebase).

3) **Git history** – To see the latest changes against the
   external code repository.

4) **Outline** – Shows the hierarchy of controls within the
   application view.

5) **Debugger** – This is to set up debugging sessions
   while developing and debugging.

6) **Test results** – It shows the unit test results.

On the bottom right of the screen, there are a few more icons that show
the console output for logs, errors, and warnings. These console outputs
are helpful when developing, building, and troubleshooting the build and
deployment of applications and services. It shows any information logged
by the system.

The **SAP Web IDE for SAP HANA** (on-premise and the SAP HANA
Express Edition, HXE) looks very close to an exact replica as shown before
and also shown in Figure 3-6. As mentioned before, the storyboard is not
present in this version; however, the Database Explorer is. So far, both
versions have pretty much the same features. One major difference in
these tools is the licensing terms. Consult the license types for these tools
on the bottom right-hand corner.

Figure 3-6. *SAP Web IDE on-premise (HANA Express Edition)*

The **SAP Web IDE personal edition**

Once again, most of the features from the SAP HANA Web IDE personal edition (Figure 3-7) have the same features present on this version of the Web IDE. One of the main differences is that the Database Explorer is not present because the personal Web IDE is primarily for the development of SAP Fiori applications and SAP Fiori extensions. In Figure 3-7, the personal edition of the SAP Web IDE can be connected to an ABAP back-end repository (via a destination file). Once connected, it can clone, pull, stage, commit, and push code via commands to the Git external repository.

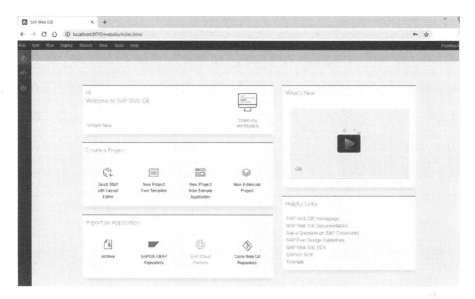

Figure 3-7. *SAP Web IDE personal edition*

Once entering the workspace section of the personal Web IDE (Figure 3-8), developers are in familiar territory again. Most of these features shown have been explained before. They work in a similar fashion as described. While the personal SAP Web IDE is installed on someone's laptop, and shown on the browser address bar, the personal Web IDE is hosted on localhost; that means that it was downloaded as an executable and configured to run from within the local development machine. Again, be sure to read the licensing terms if deciding to use the personal Web IDE for any Fiori development.

Figure 3-8. SAP Web IDE personal edition workspace

After introducing the SAP Web IDE (and its different editions), continue to see how to connect it to an external code repository, GitHub. In Chapter 1, it was briefly showcased how to set an external code repository from the Visual Code IDE. The remaining development and exercises in the book will be completed from the SAP Web IDE.

After starting the exercise with a database module, proceed to connecting the wavepress project to a GitHub repository so that it can be cloned.

In the SAP HANA Web IDE, go to the project name. Right-click, select Git, and select Initialize Local Repository (Figure 3-9). Provide an email and username on the next step that matches a Git account as shown in Figure 3-10.

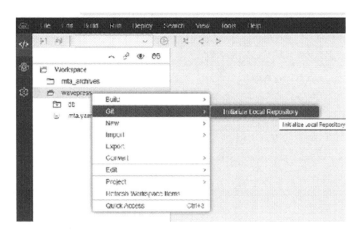

Figure 3-9. Initialize Local Repository

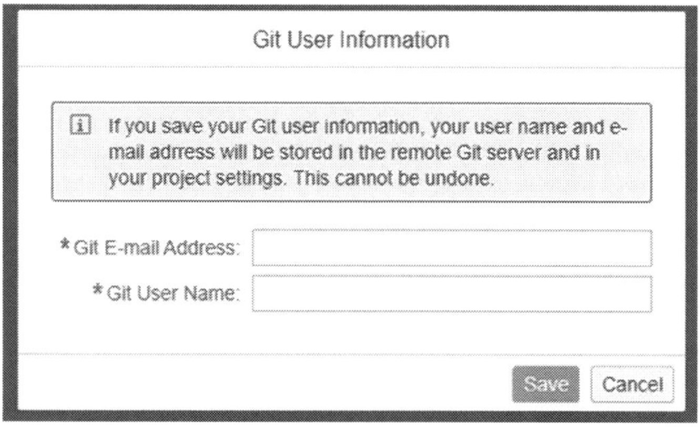

Git User Information
If you save your Git user information, your user name and e-mail adrress will be stored in the remote Git server and in your project settings. This cannot be undone.
*Git E-mail Address:
*Git User Name:
Save
Cancel

Figure 3-10. *Git User Information prompt*

After this step, a local GIT repository becomes available for development, or even better, a remote repository can be set at this time by clicking the Set Remote button from the message shown in Figure 3-11.

Rebase
Merge
Name
Stage

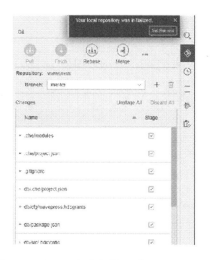

Figure 3-11. *Local repository initialized message*

If Set Remote is selected, then specify the URL of the remote repository to be connected. In GitHub, create a remote repository before connecting to it. As was demonstrated in Chapter 1, open the repository and select its URL. That URL (ending in git) will be used on the very next step from the SAP Web IDE wizard. At this point, if Git is used, it may be necessary to select "Add configuration for Gerrit" and select OK in Figure 3-12. Gerrit is a code review feature that works with Git.

Since no SSO has been set up on GitHub, authentication into the GitHub account is done from the SAP Web IDE using credentials (basic authentication).

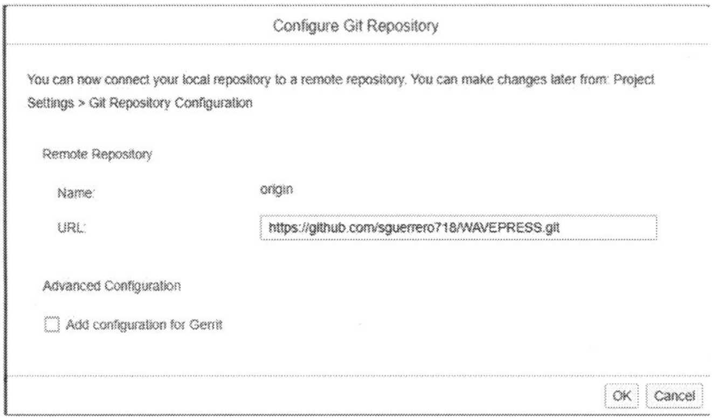

Figure 3-12. *Configuring remote repository*

Provide credentials to be authenticated into the GitHub account as shown in Figure 3-13.

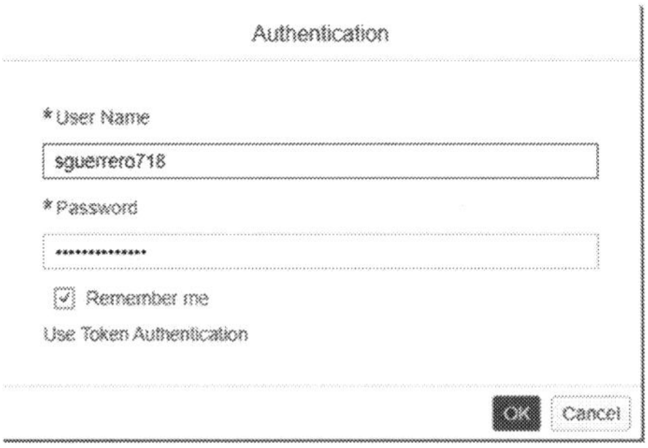

Figure 3-13. *Git prompt for authentication*

If authenticated, a message will be received to fetch the initial commit as displayed in Figure 3-14.

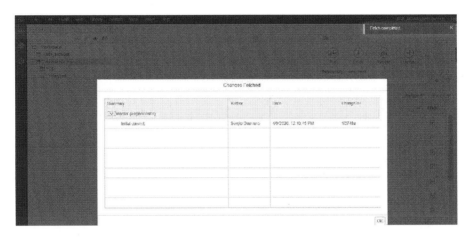

Figure 3-14. *Initial GIT fetch*

The SAP Web IDE initialized the main branch and will start tracking the changes going forward. As the project has been initially created with a database module, the additional pending changes needed to be staged, committed, and pushed as done and shown before. One of the many nice features about this tool is its ability to track pending changes. It allows developers to enter comments and, moreover, it has an interface that mimics the git commands from the command line. It is worth noting a few things that are very important in Figure 3-15:

1) The ** on the Workspace section shows that the current project is mapped to an external repository.

2) The mapped branch is the master branch. It shows square brackets on the project section, and it is also selected from the drop-down in the Git pane.

3) Any pending changes are listed on the Git pane
    (right navigation menu). These changes can be
    staged, committed, and pushed within the same
    click – do not forget to provide a meaningful
    comment; otherwise, the Commit and Push button
    will not be enabled.

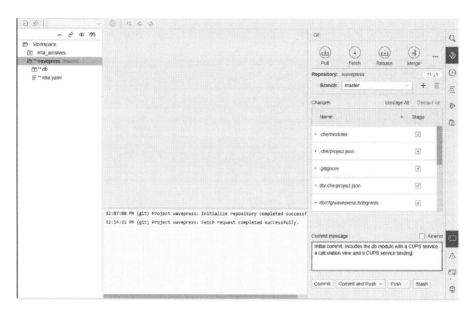

Figure 3-15. *Initial commit from the SAP Web IDE to the remote GIT repository*

Proceed with the initial commit – if an error is received, then perform a Pull first (after the initial mapping of the remote repository as shown in Figure 3-16). A Pull is initially required.

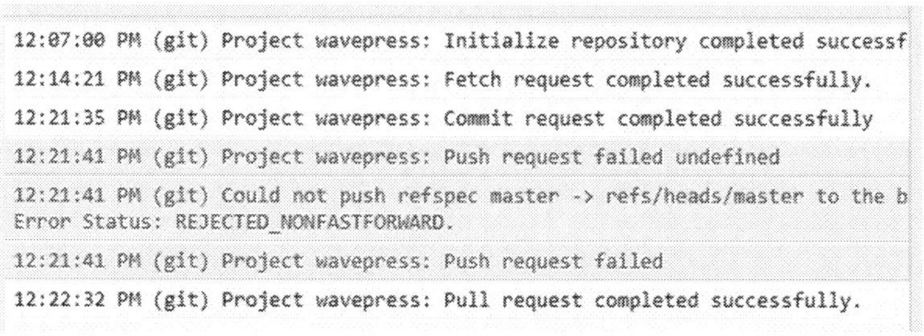
12:07:00 PM (git) Project wavepress: Initialize repository completed successf
12:14:21 PM (git) Project wavepress: Fetch request completed successfully.
12:21:35 PM (git) Project wavepress: Commit request completed successfully
12:21:41 PM (git) Project wavepress: Push request failed undefined
12:21:41 PM (git) Could not push refspec master -> refs/heads/master to the b
Error Status: REJECTED_NONFASTFORWARD.
12:21:41 PM (git) Project wavepress: Push request failed
12:22:32 PM (git) Project wavepress: Pull request completed successfully.

Figure 3-16. *GIT message from the SAP Web IDE console*

After the successful Pull, some commits may show in pending status (number with arrow up at the end of the line where **Repository:** wavepress shows in Figure 3-17). After the initial Pull is completed, proceed to run a Push command (notice that the Push button has become enabled).

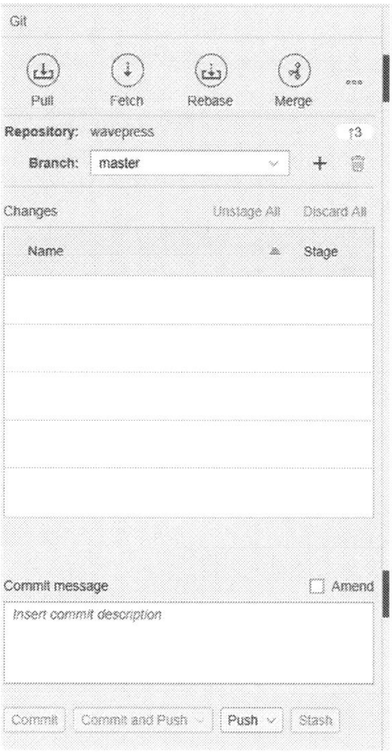
Git
Pull
Fetch
Rebase
Merge
Repository: wavepress
↑3
Branch: master
Changes
Unstage All
Discard All
Name
Stage
Commit message
Amend
Insert commit description
Commit
Commit and Push
Push
Stash

Figure 3-17. *SAP Web IDE GIT pane master branch*

From the console log, messages for the Pull and Push operations are shown as completed correctly in Figure 3-18.

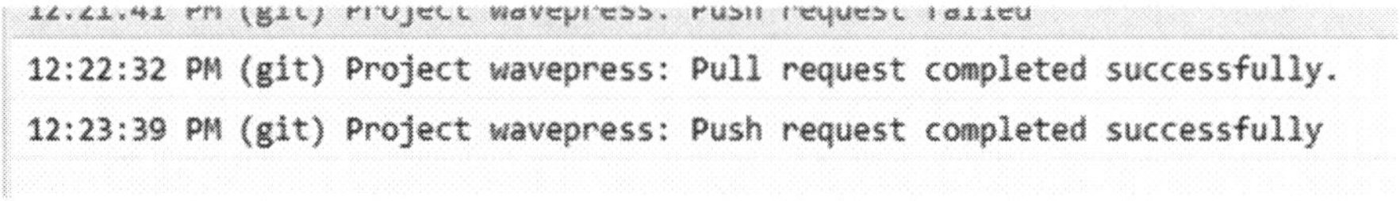

```
12:21:41 PM (git) Project wavepress: Push request failed
12:22:32 PM (git) Project wavepress: Pull request completed successfully.
12:23:39 PM (git) Project wavepress: Push request completed successfully
```

Figure 3-18. *Pull and Push request messages from the SAP Web IDE console*

Return to the GitHub repository to see and compare the repository before and after the Push operation (shown in Figures 3-19 and 3-20).

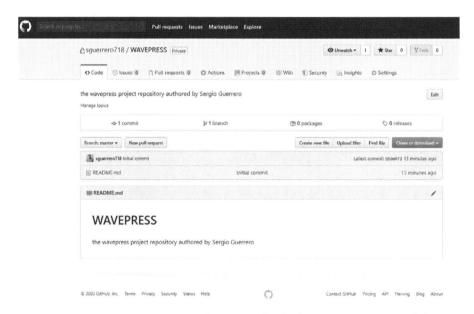

Figure 3-19. *GIT repository showing the before committing of the SAP Web IDE*

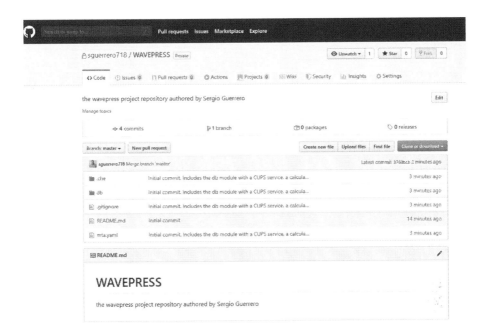

Figure 3-20. *GIT repository after pushing code from the SAP Web IDE*

In the workspace (SAP Web IDE), notice the ** symbols have also changed to a different identifier, single dot, when the code is in sync between the SAP Web IDE and the Git repository. Proceed with the following set of related changes and add additional commits to the message box (following the best practices of committing often small, related changes (Figure 3-21)).

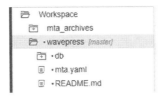

Figure 3-21. *Code files display an icon representing the tracking of changes*

Database Explorer

The Database Explorer is a tool that comes with the SAP Web IDE and
facilitates connecting to HDI containers and classic SQL schemas in a
SAP HANA system. The Database Explorer allows developers to run SQL
queries from the SQL console. It also allows viewing the various objects
such as Tables, Column Views, Procedures, Tasks, Triggers, and Synonyms.
The Database Explorer is used for development, debugging, testing,
and running commands when needing to create objects in a classic
environment schema (noncontainer, runtime-based objects). One or more
database connections can be established from the Database Explorer as
can be seen in Figure 3-22.

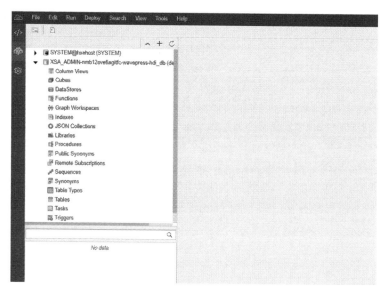

Figure 3-22. *Multiple database connections shown from the*
Database Explorer

The first and easiest way to become familiar with the Database Explorer and its features is by creating a database module from the SAP Web IDE Workspace. When visiting the Database Explorer for the first time, it will allow the creation of a connection to the HDI container (as shown in Chapter 2) or classic SQL schema. After creating a database module in the project, developers can see, query, and modify database objects belonging to the same container. In a real-world scenario (described in Chapter 2), there will be models and procedures interacting from within HDI containers and from within classic schemas. The need to create mixed data models and other database artifacts is a normal development activity. The creation of CUPS, HDB grants, and synonyms are prerequisites in situations when artifacts from HDI and classic schema objects are used. The ability to query synonyms and classic objects from the SQL console is achieved after the respective modules are built and the connection to the database container/schema has been established as shown in Figure 3-23. Here are a few queries using the DEVICES and DUMMY (Figure 3-24) synonyms.

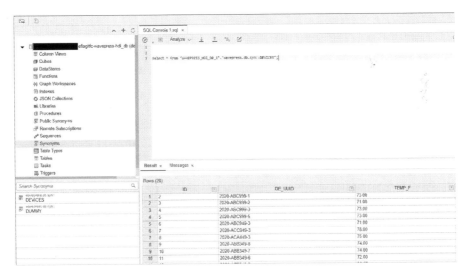

Figure 3-23. *SQL console displaying a query using a synonym*

Figure 3-24. *Synonym for the DUMMY table*

Some very important features in the Database Explorer to take advantage of when analyzing database development are the Explain Plan and the SQL Analyzer.

The Explain Plan helps by breaking down the process of running a SQL statement into the involved database objects, operators involved, connection details, and time to perform operations (Figure 3-25).

Figure 3-25. *Explain Plan*

The SQL Analyzer offers a different type of analysis.

In the **Overview** tab (Figure 3-26), the time of "Compiled vs. Executed" is shown in the first section. This section is helpful to troubleshoot database objects when troubleshooting performance.

Another section within the Overview tab is the SQL Performance Recommendations. Currently, there are none. This section may be populated when a more complex statement is analyzed.

The third section within this tab is the Dominant Operators section where performance metrics of the different nodes and execution in the database model can be seen.

Finally, the Statistics section has information from the system, the executed view, the number of records outputted, and the memory allocated during the execution of the SQL statement.

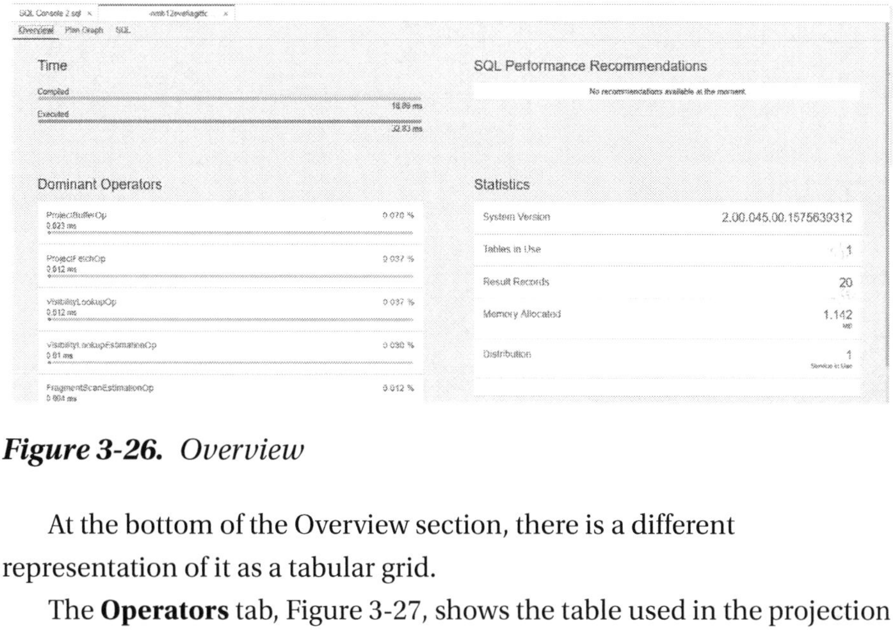

Figure 3-26. Overview

At the bottom of the Overview section, there is a different representation of it as a tabular grid.

The **Operators** tab, Figure 3-27, shows the table used in the projection and additional metrics such as execution and CPU time, to mention a few.

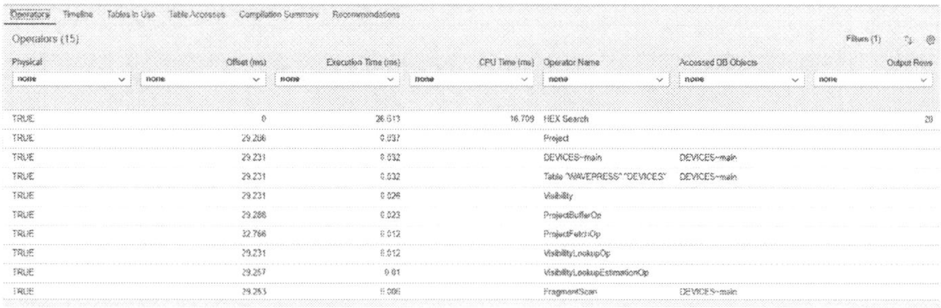

Figure 3-27. Operators

The next tab is the **Timeline** (Figure 3-28). Within this tab, the timeline incrementally shows what happens from the bottom projection up to the semantic layer in millisecond intervals. Different types of execution visuals will appear when analyzing these segments as they may be larger or smaller depending on the overall time to completion.

Figure 3-28. *Timeline*

Tables in Use is self-explanatory in Figure 3-29; the simple information model shows the projection of one table.

Figure 3-29. *Tables in Use*

In Table Accesses tab, there is only one table as shown in Figure 3-30. However, the steps and processes are shown as if they occurred during the execution of the query in context.

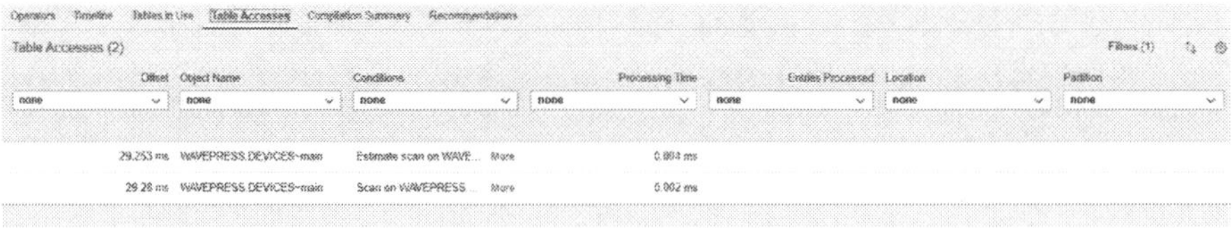

Figure 3-30. *Table Accesses*

In Compilation Summary, expand the sections to find detailed information as in Figure 3-31.

Figure 3-31. *Compilation Summary*

Compilation Summary is expanded in Figure 3-32.

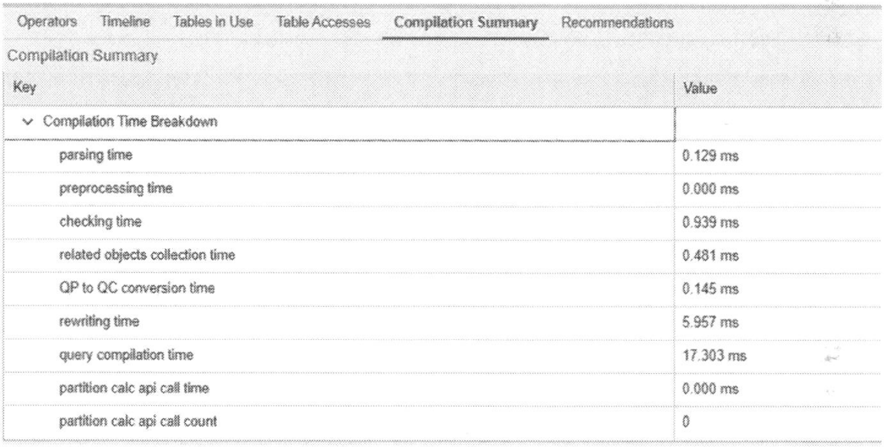

Figure 3-32. *Compilation Summary – time breakdown*

Recommendations is shown on the **Overview** tab and in Figure 3-33. There are none for this database model.

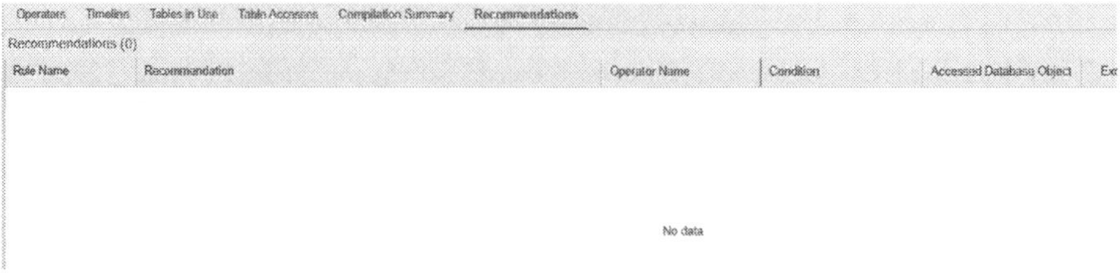

Figure 3-33. *Recommendations*

Next to the Overview tab, there is the **Plan Graph** tab in Figure 3-34. Most developers with experience in SAP HANA 1 will quickly become familiar with this section due to it being almost an exact replica of the query plan execution that exists in SAP HANA Studio. It now appears in the Database Explorer of the SAP Web IDE. The Plan Graph will show the various nodes that are being executed in the information model from top to bottom in a graphical representation. The nodes within the graph contain very helpful information such as the number of records returned from one node to the next. The **inclusive** time and incremental time are also present as execution happens from the bottom node upward. The **exclusive** time, the current operator executing time, and the time to complete a query are other metrics displayed in this representation.

Figure 3-34. *Plan Graph*

Finally, the **SQL** tab shows the executed SQL statement as shown in Figure 3-35.

```
Overview    Plan Graph    SQL

SELECT *
FROM "WAVEPRESS_HDI_DB_1"."wavepress.db.models::CV_DEVICES"
```

Figure 3-35. *SQL statement*

SAP HANA XS Advanced Cockpit

The **SAP HANA XS Advanced Cockpit** is a tool that is primarily used for SAP HANA XS advanced developer team leads and XSA system administrators as shown in Figure 3-36. This tool allows developers/administrators to navigate the XSA Organization and see the spaces, users, roles, applications, and service instances. The SAP HANA XS Advanced Cockpit landing screen looks like Figure 3-36.

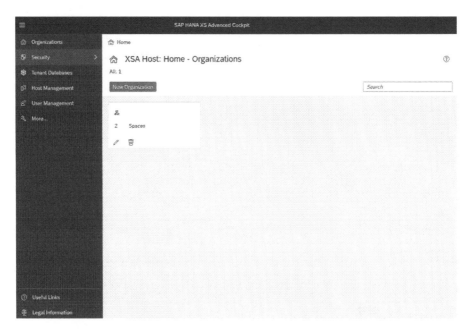

Figure 3-36. *SAP HANA XS Advanced Cockpit*

This tool can be accessible via SAP Web IDE in one of two ways: by navigating to Tools, SAP HANA XS Advanced Cockpit, as shown in Figure 3-37.

Figure 3-37. *Accessing the cockpit from the SAP Web IDE*

or from the XSA controller URL, YOUR_HOST:39030 (default port), as
shown in Figure 3-38. Be aware that other XSA applications provided by
SAP are accessible from these links as well.

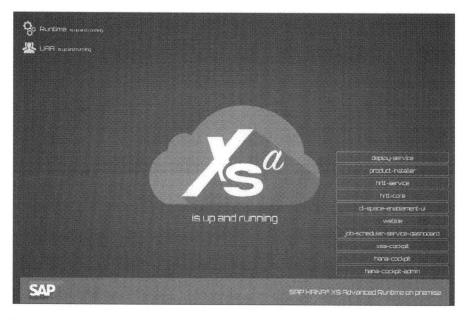

Figure 3-38. *SAP XSA controller*

As described in the Cloud Foundry section, have awareness of the CF
and XSA structure hierarchy. **Spaces** are found as part of an organization.

In a default installation of SAP HANA XSA, the ***development*** and the
SAP spaces are included as default. In Figure 3-39, the count of current

applications within the **Spaces** is shown. Information about how many resources and memory are being utilized is also displayed. An important thing to observe from the default two Spaces is that the SAP Space should not be edited. The reason for this is because SAP applications that are running in it include the SAP Web IDE, the XSA cockpit, and the Database Explorer, among all the other 34 applications/microservices that show as **Started** status.

Figure 3-39. *XSA Organization*

As mentioned in Chapter 1, inside of each Space, there are *applications* and *services.* To validate this, click the development Space and drill into that area as shown in Figure 3-40. At this point, only a database module has been created which was built with the di-builder. Therefore, it should only display one application inside the development Space. Since the application status shows a green *Started* status, that means that it is running correctly; if the status was shown as "Stopped," it means there are some issues that can be seen from the application logs.

Figure 3-40. *XSA Spaces*

Figure 3-40 is very important due to additional details within the Space are displayed:

1) **Monitoring**

    It shows the application monitoring settings such as the amount of memory consumed by the applications, the CUPS service, user mode times, the mta yaml file, and the host where it is running from.

2) **Services**

    a. Service marketplace

        These are services provided by the framework out of the box, such as the XS UAA service, the job scheduler, managed HANA db service, or SAPUI5 service, among others.

    b. Service instances (Figure 3-41)

        This one is very important because services can be created in different ways, and those service instances will be displayed here. In previous development steps, service instances were created (the UAA service instance) from the db module assigning access to a cross-schema service. During the earlier exercise, WAVEPRESS_SVC was used; however, some

attempts yielded incorrect authentication, and
consequently, those service instances are still
showing here. The cleanup of these instances
needs to be done with proper care so that the
correct instances are removed and those being
used remain untouched.

Figure 3-41. *XSA custom services*

    c.   User-provided services (UPS) (Figure 3-42)

The custom user-provided services are used
when the platform marketplace does not offer
such services. In the current example, a UPS
service for cross-schema access was created.
Opening the details of the UPS, its content can
be seen. Be extremely careful when granting
access to the cockpit application since users
with this access will be able to see the raw
connection details (including credentials for
the schema) for this UPS service instance. Only
authorized users to sensitive system credentials
should have access to the XSA cockpit.

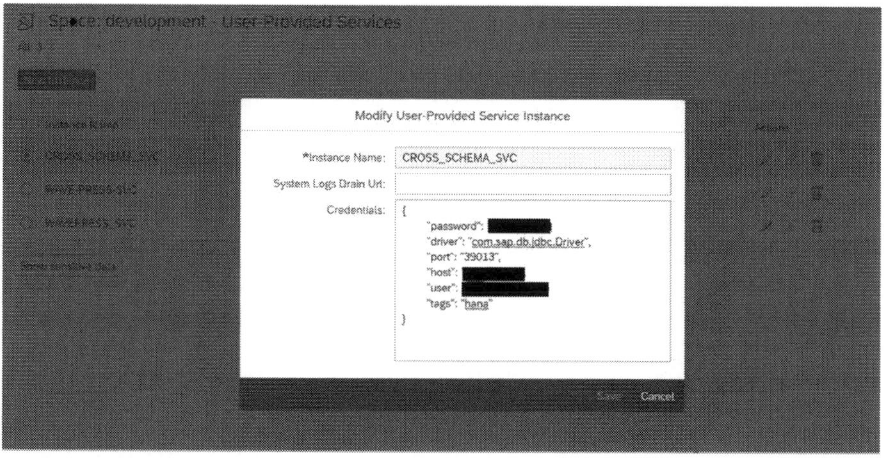

Figure 3-42. Custom service connection details

3) **Routes** (Figure 3-43)

Currently, there is only one route related to the di-builder application. Later, when creating the REST API, additional routes will be shown in the Routes section.

Figure 3-43. XSA routes

4) **Members** (Figure 3-44)

This section shows users mapped to the current Space. In the default setting of the development Space, the initial two users and their status within the Space are shown. Use these accounts as templates for other users that need to be created.

111

Figure 3-44. *XSA Space Members (users)*

5) **Pinned hosts** (Figure 3-45)

   Currently, no Space is pinned to a host.

Figure 3-45. *Cockpit Pinned Hosts*

Analyze the different features within the XSA cockpit shown in Figure 3-46. The first section is **Security**. This section allows us to see and manage the existing Role Collections (groups of roles). The default configuration of an XSA environment includes some role collections for the XSA developer and the XSA administrator as shown in Figure 3-46.

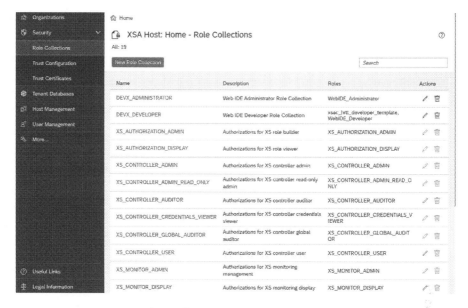

Figure 3-46. *XSA Role Collections*

The **Trust Configuration** (Figures 3-47 and 3-48) section allows administrators to set up a trust connection between the XSA environment and external systems. Settings like single sign-on are configured here. Anything requiring a trust can be set up from the New Trust Configuration button shown next.

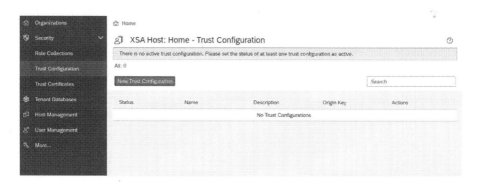

Figure 3-47. *XSA Trust Configuration*

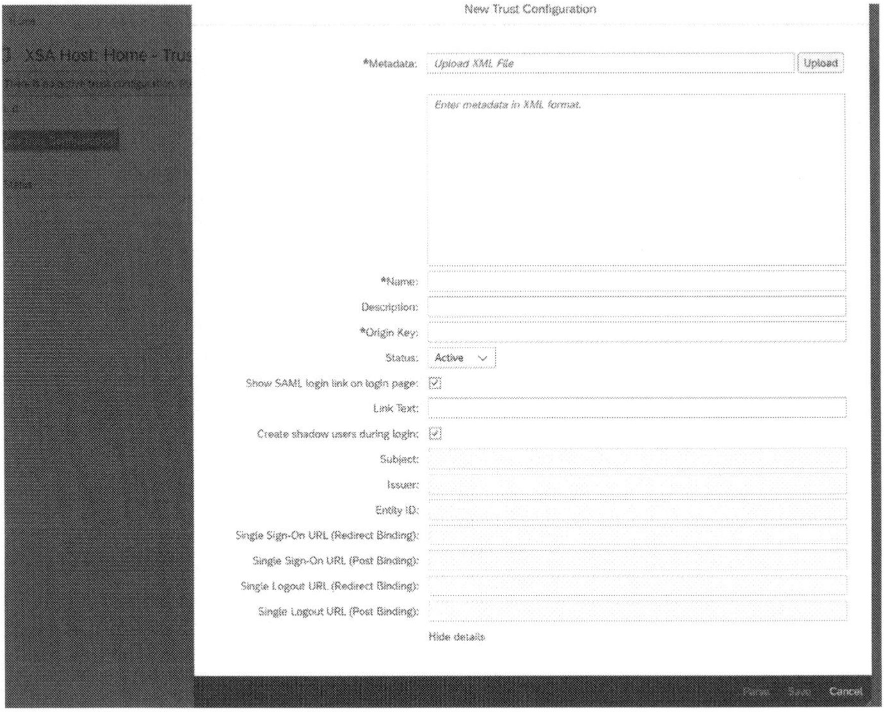

Figure 3-48. *Setting a new Trust Configuration*

Trust Certificates allows users to set up trust certificates to connect to other systems using a specified certificate (either a domain specific or one signed by a global authority CA) as shown in Figure 3-49. Alternatively, the XS CLI can be used to set the cert using the **xs set-certificate** command.

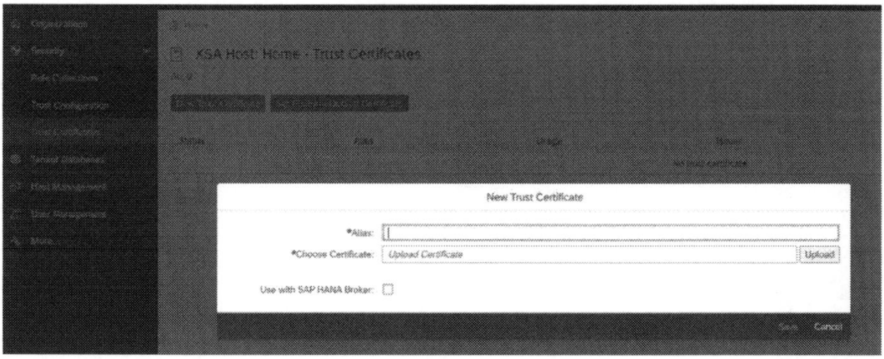

Figure 3-49. *Setting New Trust Certificate*

Under Security in the XS Advanced Cockpit, the **Tenant Databases** item is shown in Figure 3-50.

Figure 3-50. *XSA cockpit Tenant Databases*

The **Host Management** section displays one or more hosts in the XSA environment as shown in Figure 3-51.

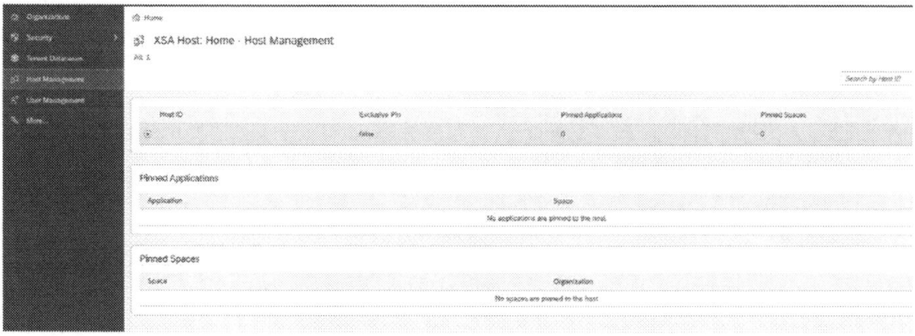

Figure 3-51. *XSA cockpit Host Management*

115

Within the **User Management** section, default users and role collections can be found from the default installation. Follow the pattern of these role collections to create additional roles for the XSA environment. The two default users are also created when installing the XSA environment as shown in Figure 3-52.

Figure 3-52. *XSA User Management*

Additional users can be created by clicking the **New User** button and providing the following details. The easiest and best approach when trying to create users in the XSA environment is shown in Figure 3-53.

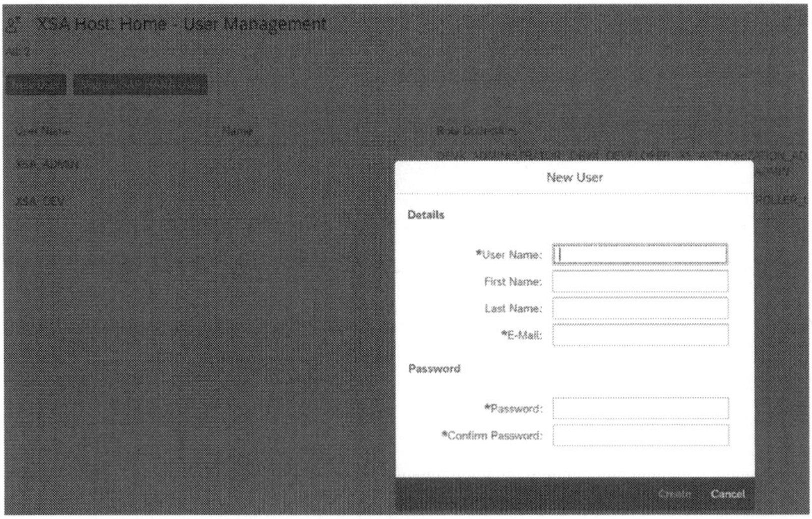

Figure 3-53. *Creating a New User*

An alternative to creating XSA users from the XSA cockpit is to **Migrate SAP HANA User** as shown in Figure 3-54. Migration of SAP HANA users must be carefully planned and executed as some properties of these users may not always migrate correctly. If an organization decides to follow this approach, carefully validate the user creation and the user access on the new HANA XSA environment.

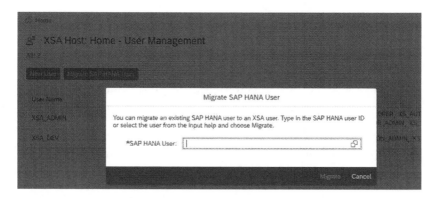

Figure 3-54. *XSA migrating a SAP HANA User*

It is important to highlight two links from the landing page of the SAP HANA XSA Cockpit:

a) **Useful links** – This will redirect to documentation related to this section and developer tutorials. It is worth navigating and spending some time in this link to understand the tool.

b) **Legal information** – This pertains to licensing terms.

POSTMAN REST client

The POSTMAN REST client is a tool that offers developers the simulation of REST API request scenarios without having to build out full-fledged applications or APIs. The POSTMAN REST client tool is a free tool that can be downloaded from any browser. There are other similar tools that may allow similar tasks to be performed (e.g., make REST API requests). However, they may have additional licensing and pricing requirements. For demonstration purposes, the POSTMAN REST client will be used as the tool of choice to demonstrate the development and unit testing of this API.

On most software systems and programming languages there are consoles, browsers, or within the IDEs, developers can perform unit tests and validate their development. On some software scenarios, additional hardware or software is required or suggested. When developing REST APIs, developers require testing HTTP methods such as GET, POST, PUT, and DELETE. Each of these HTTP methods requires different levels of testing. The easiest HTTP method to test is the GET method. Its request only reads data from a system. Developers can easily accomplish a GET request from a browser URL while having an Internet connection (if accessing a service outside the network). The other HTTP methods are a little bit trickier to test since anything altering the state of the system (PUT, POST, DELETE) requires an additional level of security provided by the consumer of the API before interacting with the back-end system – the OAuth token.

In Chapter 1, section 1.2, it was mentioned that there are some security considerations when requesting access to a resource. Moreover, certain protocols used by software systems and industry standards are enforced to protect the integrity of such software systems and sensitivity of data. Also, in Chapter 1, section 1.2, a managed service called UAA was mentioned. It was also explained how it uses OAuth 2.0 to accept a request, analyze its parts, and either validate it and return an access token or deny access to the back-end resources.

Additional features of the POSTMAN REST client tool allow the creation of workspaces to separate unit tests, manual testing and automated testing. It allows the integration into continuous integration and delivery pipelines using POSTMAN API commands. Visit the POSTMAN website for additional rules and pricing details.

What needs to be done to start POSTMAN REST client testing?

First, download the tool from the browser and accept the licensing terms – as it applies to the version supported by an organization/team that will be using it for development and validation. Then, break the testing down into how it will be used in this tool and in their development. Separate the API calls based on how logical units of work may seem appropriate, for example, by environment, by endpoint, by APIs (if you have more than one).

Begin by testing out some free online API endpoints to showcase this part of the tool. When getting into the exercise API, Chapter 4 will showcase the POSTMAN request from the actual project that will be built.

Open the tool and click **New.**

Provide details for a request and select (or create) a collection to save the request. Collections resemble folders for POSTMAN organization of requests. The collection shown here is called TEST API and contains one request called TEST API (Figure 3-55). Within the request, add details such as Method=**GET**, URL = {a free URL found online}, and, if no other request parameters need to be set, such as request header or authentication, send the request.

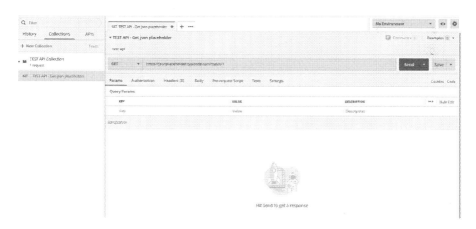

Figure 3-55. POSTMAN

After clicking Send, see several important details:

1) Figure 3-56 shows the response (bottom half of the screen) containing the Status (200), the Time it took to receive a response (39ms), the Size of the payload (689B), the response in JSON format, a cookie, and some Response Headers. This is a very basic request, but it shows most of the details of the request and the response from the POSTMAN tool toward an external service.

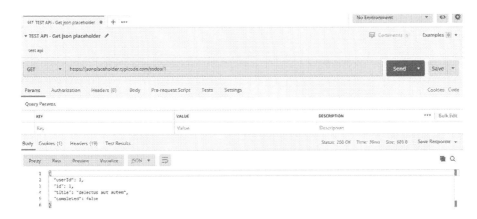

Figure 3-56. *Sample POSTMAN request*

To see the same request from another tool to verify the POSTMAN request, copy the same URL into a browser window and send it. If this example is used for practice, passing additional parameters will be required due to authentication (inside the Authorization tab) and perhaps some request headers (Headers tab) to send and receive JSON formatted data. The output of the request can be seen in Figure 3-57.

Figure 3-57. *Free REST API and Network tab on the browser*

While the same request can be validated from two different tools, going forward only the POSTMAN REST client tool will be shown on the actual project development. It is important to note that POST, PUT, and DELETE cannot be simulated from the browser; only GET requests can be validated from the browser address bar.

Conclusion

At the end of this chapter, you should be able to understand the different variations of the SAP Web IDE as their primary tool for the development of microservices within the SAP HANA XSA environment. There are other tools that are used to keep our software synched and stored in an external code repository. As development teams grow, these code repository tools are needed to be able to have code fixes and new feature development progressing in parallel. Additional troubleshooting capabilities were highlighted from the Database Explorer to address issues with performance on database models and stored procedures.

CHAPTER 4

SAP HANA XSA NodeJS development

This is probably the chapter everyone is waiting for. In this chapter, the book will explain the Node JS development as it applies to SAP HANA XSA. There are various endpoints and also HTTP methods discussed in detail to handle the request and response from the microservices exercise being built. Fasten your belts and let's get going.

Development of REST APIs

In Chapter 1, the definition of REST APIs was introduced. One of the most important factors to successful development is understanding the differences of REST APIs and when and how to use them. REST APIs were created to expose and ingest data via HTTP(s) methods (Table 4-1). Another benefit of using REST APIs is to allow consumers of these APIs to get access to data from businesses without having direct access to the database. The consumer of the REST API must know the URL to consume the API, the required authentication, the expected data structure to be provided to the API, any response structure to be returned from the API, and possible error messages.

© Sergio Guerrero 2020
S. Guerrero, *Microservices in SAP HANA XSA*, https://doi.org/10.1007/978-1-4842-6118-7_4

Table 4-1. *HTTP methods*

HTTP method	When to use it
GET	Read data
POST/PUT	Insert/update data
DELETE	Yes, you guessed it right
OPTIONS	Preflight options – the server will respond whether it is acceptable to send a request with these parameters

Most REST implementations begin with the use of the GET method (also described and shown before from the POSTMAN section). In this section, start by creating the Node JS module in the HANA XSA application. Open the SAP Web IDE. Keep an eye on the mta.yaml file because as additional modules are created or updates are made to module dependencies, this file may require manual intervention. So far, there is only a module for the database development (**db** module). However, since starting to work with a back-end module, it is important to understand the module dependency and why and where to make the next set of updates.

From the SAP Web IDE, right-click the project, select New, and then Node.js Module as shown in Figure 4-1.

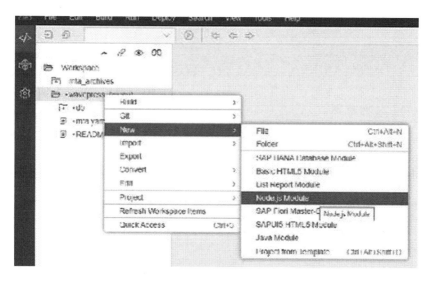

Figure 4-1. *Creating a Node.js Module*

After making this selection, follow the wizard steps. Call the Node JS module **api** since it will represent the REST API (Figure 4-2).

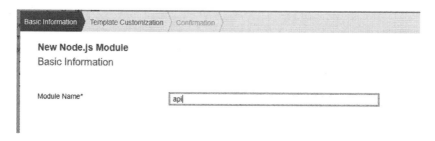

Figure 4-2. *Module name*

Continue providing the module settings (such as version 1.0.0) since it is the initial step and enable XSJS support since it was mentioned on the previous subsection and displayed in Figure 4-3.

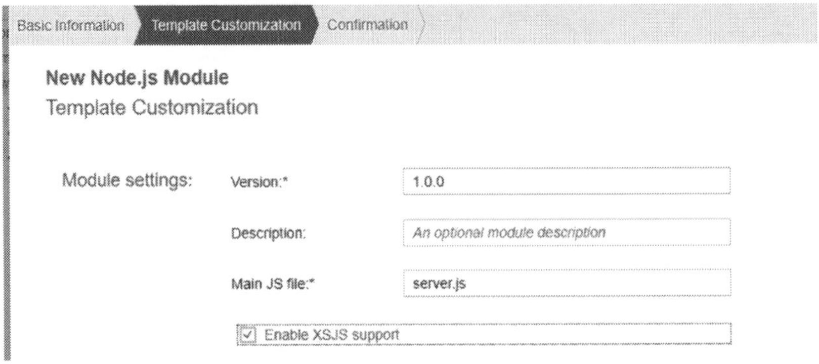

Figure 4-3. Node.js Module settings

Click **Next** to confirm or simply click **Finish** as shown in Figure 4-4.

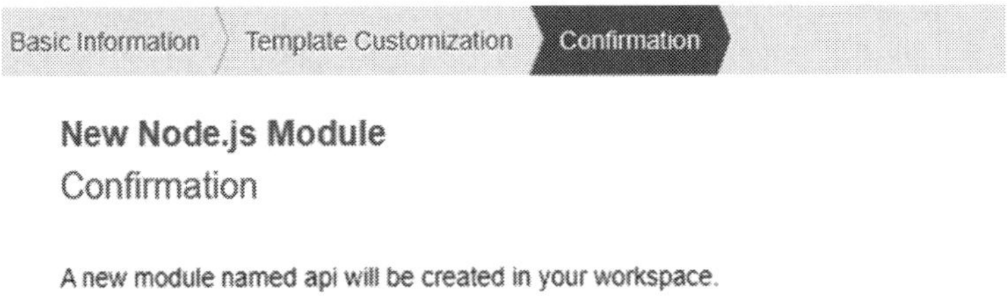

Figure 4-4. Confirmation of module creation

After the wizard completes this step, the **api** module that was created and added to the project can be seen from the Web IDE. Additionally, opening the mta.yaml file will display the new entry of this module in the wavepress application as shown in Figure 4-5.

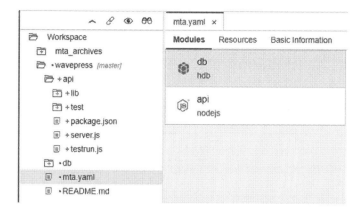

Figure 4-5. *mta yaml file after module creation*

Now, there is a node module with some default settings such as the XSJS compatibility mode and some default files created by the IDE.

In most software development involving Node JS, a package.json file will also be found. This file contains information about the node dependencies, the metadata of the module, what file to run on start of the module, and any unit test files that could be set up during the project development as shown in Figure 4-6. This file would only be updated if there are additional dependencies included or removed.

```
package.json  x
 1 ▾ {
 2 ▾     "dependencies": {
 3           "@sap/xsenv": "^2.0.0",
 4           "@sap/xsjs": "^5.2.0"
 5       },
 6       "description": "my description",
 7 ▾     "devDependencies": {
 8           "@sap/xsjs-test": "^3.0.2"
 9       },
10       "files": [],
11       "main": "server.js",
12       "name": "api",
13 ▾     "scripts": {
14           "start": "node server.js",
15           "test": "node testrun.js"
16       },
17 ▾     "engines": {
18           "node": "8.x"
19       },
20       "version": "1.0.0"
21   }
```

Figure 4-6. *package.json file within Node.js module*

Currently, the default Node JS module is a basic file that can be run with no changes in it; the module must be built and run. It is recommended to build the module before proceeding with other development to ensure the current module can be built and that any dependencies can be downloaded from the npm package manager. Some of the dependencies downloaded will be open source, and some are going to be @sap modules.

Since it is the first time building this module (Figure 4-7), it may take a few seconds to download the dependencies and create the node application in the system. If the download of the npm modules fails, analyze the console window to see messages such as a version of a module is not correct or found or maybe see if the npm site is reachable. The SAP Web IDE is attempting to download some modules from the Internet, and it will require access to external URLs.

Figure 4-7. *Module build*

Receiving a successful message here means that the module was built successfully. If an error is received, then it must be analyzed before proceeding. Some errors that may occur at this point may be due to npm module dependency versions not found or npm modules not downloaded. Read the console output errors in order to see what may have gone wrong. After making changes on npm modules or versions, repeat the rebuild process to get a successful build before proceeding. If the XSA environment is hosted on a cloud provider, chances are the environment may require additional permissions to be able to connect and download content from npm or an external package manager. If these issues occur, contact the environment administrator to ensure proper connectivity to external resources can be done.

If the build is successful, the following message will appear with a "completed successfully" message in the Web IDE console as shown in Figure 4-8.

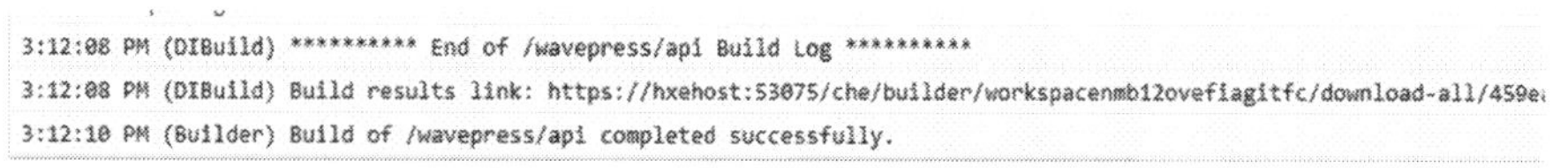

```
3:12:08 PM (DIBuild) ********** End of /wavepress/api Build Log **********
3:12:08 PM (DIBuild) Build results link: https://hxehost:53075/che/builder/workspacenmb12oveflagitfc/download-all/459e:
3:12:10 PM (Builder) Build of /wavepress/api completed successfully.
```

Figure 4-8. *Console output after module build*

Run the api module by right-clicking it, select Run, then select Run as Node.js Application. The SAP Web IDE will attempt to run the application as shown in Figure 4-9.

Figure 4-9. *Running the Node.js application*

If it runs, the console window will display which URI location this module is running from – host:51026 – as shown in Figure 4-10.

Figure 4-10. *Console output after building and running the Node.js module*

After the module starts running, the console window will display some messages. Where and how was this output generated?

Within the node module, there is a file named package.json. This is a json file used during the build and run of the Node JS application. Within the package file, there is a section containing a *scripts* folder that further contains a command called *start*. This command invokes the **server.js** file.

That server.js file, shown in Figure 4-11, is the entry point of the node module in the REST API. See the console log statements in the figure at lines 18, 25, and 31. These statements generate the output to the console window within the SAP Web IDE.

```
package.json  ×    server.js  ×
1   /*eslint no-console: 0, no-unused-vars: 0*/
2   "use strict";
3
4   var xsjs  = require("@sap/xsjs");
5   var xsenv = require("@sap/xsenv");
6   var port  = process.env.PORT || 3000;
7
8 ▾ var options = {
9       anonymous : true, // remove to authenticate calls
10      auditLog : { logToConsole: true }, // change to auditlog service for productive scenarios
11      redirectUrl : "/index.xsjs"
12  };
13
14  // configure HANA
15 ▾ try {
16      options = Object.assign(options, xsenv.getServices({ hana: {tag: "hana"} }));
17 ▾ } catch (err) {
18      console.log("[WARN]", err.message);
19  }
20
21  // configure UAA
22 ▾ try {
23      options = Object.assign(options, xsenv.getServices({ uaa: {tag: "xsuaa"} }));
24 ▾ } catch (err) {
25      console.log("[WARN]", err.message);
26  }
27
28  // start server
29  xsjs(options).listen(port);
30
31  console.log("Server listening on port %d", port);
```

Figure 4-11. *server.js file*

It is important to examine this file in more detail going forward.

The initial lines 4 and 5 are requiring dependent node modules provided by the SAP registry.

The options object starting in line 8 says that authentication will not be used (anonymous: true), and eventually it will be redirected to the /index. xsjs file. At this point in development, the compatibility mode can be seen in action.

A few other options get assigned while configuring SAP HANA on lines 15–19 and the UAA service on lines 22–26. Remember that there are no UAA service instances configured yet, and therefore there will be no authentication prompt for the URL that is mapped to the api module when it is launched.

On line 29, the console output shows what was originally received when running the module.

As mentioned, the host and port listed on the console is the port bound where the node application is running. Clicking it will launch the back-end REST API running as the api module, and the redirection to the XSJS compatibility mode is seen from line 11. The output from the index. xsjs file is shown in Figure 4-12.

Figure 4-12. *Hello World output from the browser*

The XSJS code that generated that output looks like Figure 4-13.

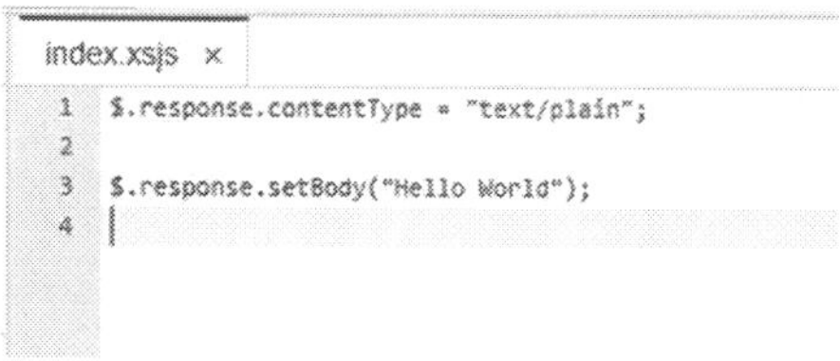

```
index.xsjs  ×
1  $.response.contentType = "text/plain";
2
3  $.response.setBody("Hello World");
4  |
```

Figure 4-13. *XSJS code*

As this node module is being developed, it may be helpful to see this application running from a back-end system (or admin view). This is useful, for example, to analyze how memory is assigned or consumed.

Using an account with the proper credentials, access this view and open the SAP HANA XS Advanced Cockpit by going to the SAP Web IDE Tools menu and selecting the SAP HANA XS Advanced Cockpit option. Figure 4-14 shows the landing page for the cockpit. Navigate to the applications by remembering the CF/XS advanced architecture hierarchy.

From the XSA Cockpit, start from the Organization and navigate to the Space (Figure 4-14). Find the microservices running under this space (as applications in Figures 4-15 and 4-16).

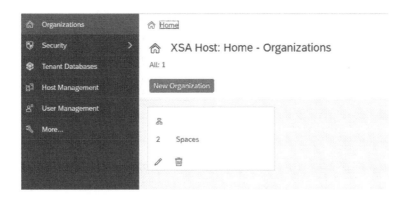

Figure 4-14. *SAP HANA XSA Cockpit*

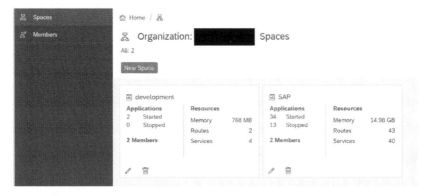

Figure 4-15. *SAP HANA XSA Spaces*

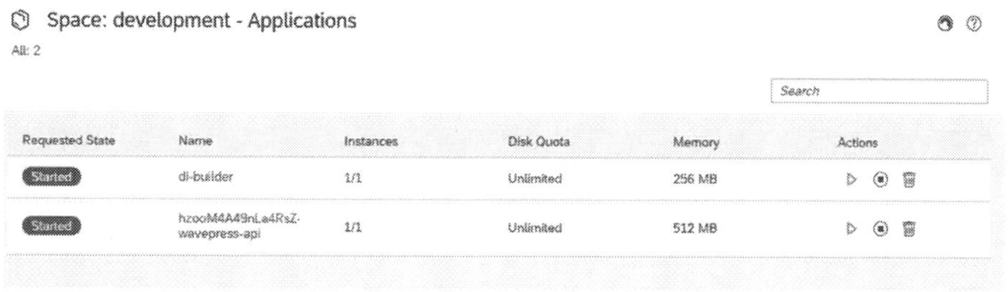

Figure 4-16. *Applications inside the XSA Cockpit*

How does validation occur if the deployed application is a microservice?

Validation happens from the initial build (of only having a database module) and now with the creation of an additional microservice (the api module). This is true as multiple processes are listed. Further, each (microservice) process can be scaled, deployed, started, or stopped independently from one another as shown in Figure 4-17.

When drilling down into the application, the following can be seen. Notice a familiar port shown on the console output; it is shown from this screen as well (port 51026).

Concepts from CF and XS will start to resemble each other as we continue to build more of the application.

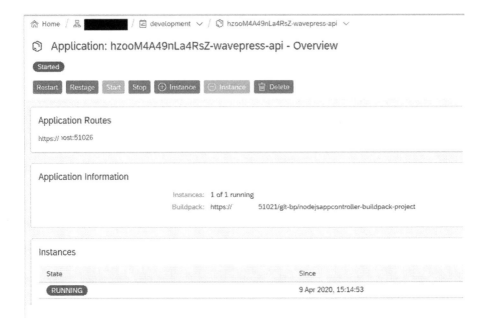

Figure 4-17. *Application settings shown in the XSA cockpit*

For the time being, leave the cockpit and return to the SAP Web IDE to continue building on the API. Keep the XSJS code where it is and continue building on the Node JS side going forward.

Recall that building an API allows interaction with a SAP HANA database table and expose data via calling a SQL script query. Saving and deleting data from the REST API will be achieved by calling a stored procedure showcasing different integration approaches.

Compatibility mode with XSJS

The compatibility mode of the SAP HANA XSA environment allows existing customers who have XSJS development from HANA 1 and are not ready to migrate their development into the HANA 2 advanced architecture to continue to run and support applications as it has been done in the XS classic architecture. The XS applications used in SAP HANA 1 are those where the SAP HANA database and XS applications coexisted together within the same environment. Moreover, the Cloud Foundry principles do not apply to these applications. This type of scenario will be briefly explained, however, for the context of the book, focus on creating REST APIs using the XS advanced architecture and the Node JS language supported from XS advanced will be shown.

It is important to mention feature differences between the two types of architectures, as shown in Table 4-2, so that if it is required to analyze and compare the current state of a company's IT infrastructure to its future, an informed decision can be made.

Table 4-2. *XSJS compatibility mode vs. XSA*

XSJS compatibility mode	XS advanced applications (NodeJS, Java, Python)
Single-thread applications (and monolithic)	Microservices that can run independently
Its performance depends on the entire system processes	Can scale (up and down) independently
Uses XSJS and $. APIs	Can leverage Node JS, Python, and other modules from external package managers and other programming languages due to BYOL (bring your own language)
Uses the HANA Studio or the SAP Web IDE for development	Can use the SAP Web IDE or other external IDEs with additional plugins
Synchronous programming	Async, nonblocking programming
OData V2	OData V4

How does the XSJS compatibility mode become enabled? It is very simple. When creating a Node JS module, a checkbox was displayed by the wizard that needed to be selected while creating the Node JS module as it is shown in Figure 4-18.

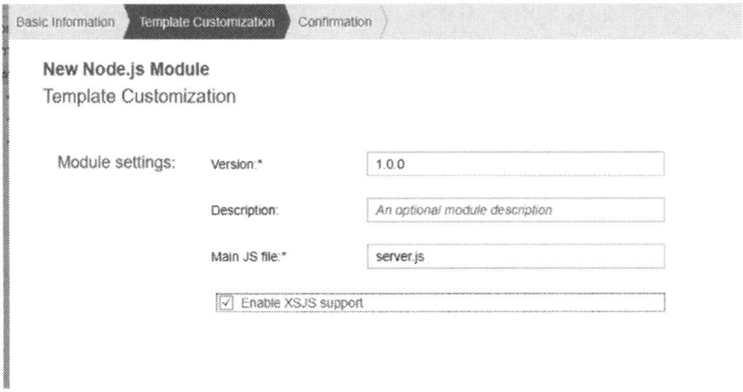

Figure 4-18. *Enabling XSJS support when creating Node JS module*

Follow the wizard steps to complete the module creation and look for XSJS files to interact with the compatibility mode in the XSA environment. The starting XSJS file for the compatibility mode is located at api/lib/index. xsjs as shown in Figure 4-19.

Figure 4-19. *XSJS file used in the compatibility mode*

Node dependencies

As building the API progresses, proven Node JS frameworks such as Express will be used for creating the API endpoints. The Express framework was described earlier in Chapter 1, section 1.2. This is one of the preferred frameworks because it is easy to use, it has great community adoption (from within and outside of SAP), the documentation is clear, and it is easy to create REST APIs in Node JS development.

To continue with the development, within the package.json file, add the dependencies for the Express framework. Other dependencies needed are the HANA database module that queries will run against, the XS environment, the XS security, logging for any issues, and the Passport module that will be leveraged while using the UAA authentication and token generation.

The package file now looks like Figure 4-20.

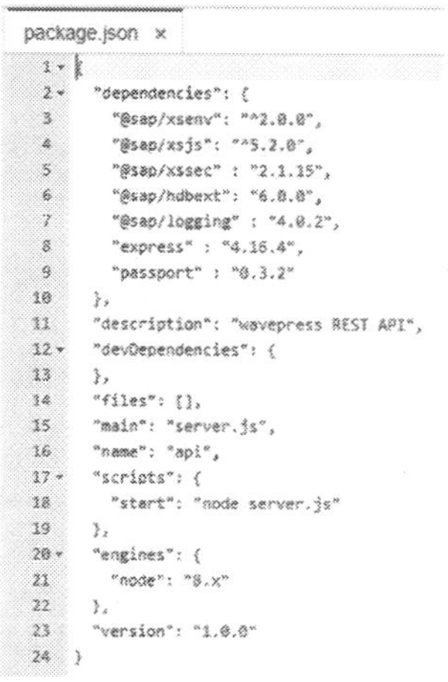

Figure 4-20. *package.json file inside the Node.js module*

In the server.js file, proceed with making a few updates to initialize Express, configure the middleware, be able to pass arguments into the framework, and make the request to a Node JS service. Following modularization and best practices, it was decided to split the routes (that will come from the request) and the service implementation as separate modules. This section of code is commonly used by SAP, and it also appears on some of the SAP tutorial sites as shown in Figure 4-21.

```
server.js  ×
1   /*eslint no-console: 0, no-unused-vars: 0*/
2   "use strict";
3   var https = require("https");
4   var xsenv = require("@sap/xsenv");
5
6   var port = process.env.PORT || 3000;
7   var server = require("http").createServer();
8
9   https.globalAgent.options.ca = xsenv.loadCertificates();
10
11  global.__base = __dirname + "/";
12  var init = require(global.__base + "init/initialize");
13
14  //Initialize Express App for XSA UAA and HDBEXT Middleware
15  var app = init.initExpress();
16
17  //Setup router (and handle its routes) & pass express app initialized
18  var router = require("./router")(app);
19
20  //Initialize the XSJS Compatibility Layer
21  //init.initXSJS(app);
22
23  //Start the Server
24  server.on("request", app);
25  server.listen(port, function() {
26      console.info("HTTP Server: " + server.address().port);
27  });
```

Figure 4-21. *Sample code from SAP's tutorials*

Notice in the server.js file, the Express module was initialized. Along with the Express module, other node and SAP modules are included. When initializing the Express module, settings related to the HANA XSA environment and the HDB client, among others, get initialized. After initializing Express, it is then passed as an argument to a router (line 18).

The router (process to handle the routes from the xs-app.json file) will be used to handle the API request on the server side, and the execution is passed from the Express framework to a specific Node JS file or function that handles any JavaScript logic, and it is executed to process the request, validate business logic, call any of the database artifacts, and eventually return a response to the requestor.

Anonymous authentication has been removed to enforce the authentication process using the UAA service. There are three types of authentication in XSA: none, basic, and route. None means that the API will be wide open and unsecure to attack. Basic means that a user provides username and password. Basic authentication is OK for simple authentication; however, it is not as robust as route authentication.

Route authentication is the preferred method as it follows best practices applied to the UAA service and the JSON Web Token (JWT). The UAA approach will be demonstrated from the api module. Authentication via the UAA service is part of the CF architecture, and that is the main reason it will be shown here.

Proceed to build the api module again and see if any errors related to any of the additional dependencies are displayed in Figure 4-22. Ensure the module builds correctly before proceeding to the next step.

If any errors related to the dependencies are shown, ensure that the module version is correct either by visiting the npm site or by ensuring the package.json file has the correct syntax. Ensure that the SAP Web IDE can connect to npm. The help of the environment administrator may be needed here.

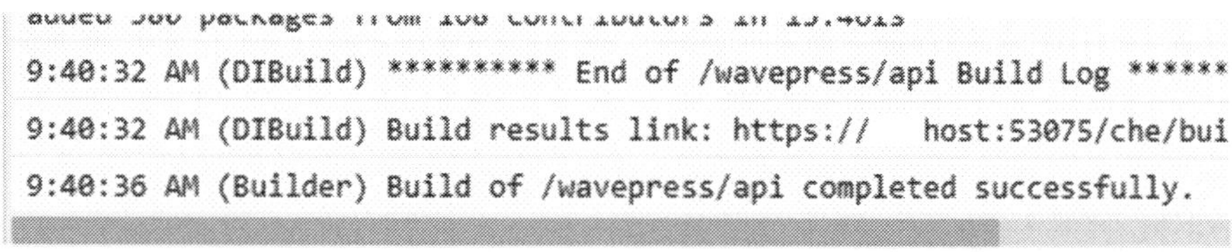

```
added 300 packages from 100 contributors in 13.4013
9:40:32 AM (DIBuild) ********** End of /wavepress/api Build Log ******
9:40:32 AM (DIBuild) Build results link: https://   host:53075/che/bui
9:40:36 AM (Builder) Build of /wavepress/api completed successfully.
```

Figure 4-22. *Console output after building the Node.js module*

139

After building the module, ensure the specified node modules were correctly downloaded. Open the application's api module and validate that the node dependencies exist inside the *api/node_modules* folder as shown in Figure 4-23.

Every time a build of this module is triggered and successfully completed, the dependencies listed inside the package.json file will be used to download or update any dependencies specified (and recursively their child dependencies as well). Each downloaded module contains its own package.json file that contains that module's own dependencies, metadata, and additional properties related to that module. Open a module's package.json file to see its version download and to make sure it matches the specified version from the api module's package json.

The node_modules folder should look like Figure 4-23 (the image was trimmed to save space – there are approximately 30–40 non-sap modules).

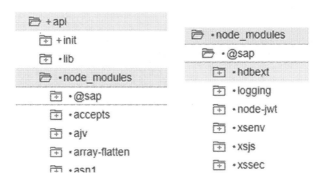

Figure 4-23. *node_modules added in the package file and downloaded into the project*

After the module is built, run it and carefully observe what happens next. Right-click the api module and select Run.

After a few seconds, the console message states that the module is running. Displayed in the console is the same host:51026 as before; however, if the URL is clicked, the browser window will display a runtime error message shown in Figure 4-24.

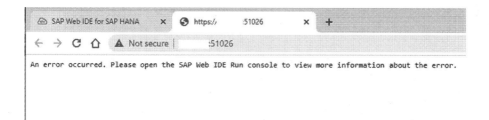

Figure 4-24. *Runtime error on the browser after enabling the UAA check during module initialization*

While this message may give an incorrect sense of whether or not the module is running due to the message shown, validate the Requested State of the api module by opening the XSA cockpit and navigate to the applications page. Notice that the applications are in fact running as shown in Figure 4-25.

◎ Space: development - Applications

All: 2

Search

Requested State	Name	Instances	Disk Quota	Memory	Actions
Started	di-builder	1/1	Unlimited	256 MB	▷ ◉ 🗑
Started	hzooM4A49nLa4RsZ-wavepress-api	1/1	Unlimited	512 MB	▷ ◉ 🗑

Figure 4-25. *View of applications running from the XSA cockpit*

This is a correct implementation, and the message shown is valid. Since the api will be utilizing UAA authentication, it will further require route authentication, and navigation routes must be declared. (Navigation) Routes are declared inside the UI module.

Create a new module and declare the dependencies between these two modules (within the mta yaml).

Right-click the project, select New and then SAPUI5 HTML5 Module and follow the prompts to add it to the wavepress project as shown in Figure 4-26.

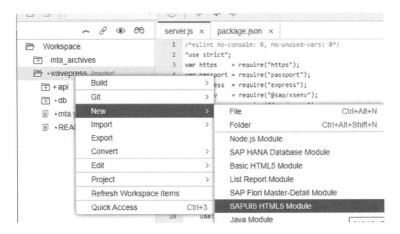

Figure 4-26. *Creating a SAPUI5 HTML5 module from the SAP Web IDE*

Add a module name. Remember when adding the other modules, this module section is also added in the mta yaml file. Shortly after, this module will need to be added as one of the api module dependencies so that it can provide route navigation to the api back end.

The namespace is useful if a front-end user interface will be built. The name of the project is used as the namespace in this example, as shown in Figure 4-27.

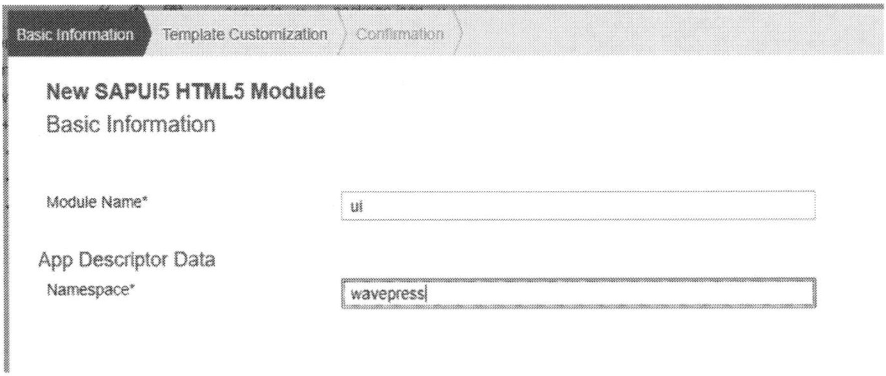

New SAPUI5 HTML5 Module

Basic Information

Module Name* | ui

App Descriptor Data

Namespace* | wavepress

Figure 4-27. *Namespace prompt after creating the HTML5 module*

SAPUI5 is a modern UI framework that uses the MVC (model-view-controller) software design pattern. When creating this module from the SAP Web IDE, provide the name of one of the UI (XML) views. By default, the wizard assigns View1 as its name, but it can be changed to another name if needed. JavaScript and HTML view types are allowed; however, SAP suggests XML views as these types of views follow a structured hierarchy, making them easier to visualize and represent from a graphical editor (Figure 4-28).

After this step, click Next and Confirm or Finish.

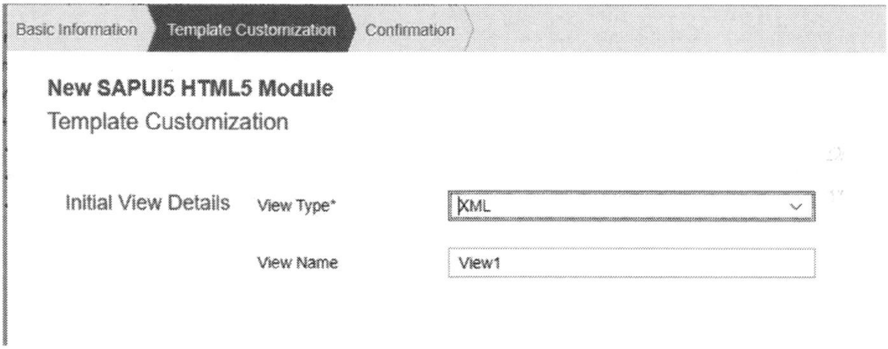

Figure 4-28. *SAPUI5 View Name prompt*

When the new HTML5 module is added to the project, it will look like Figure 4-29. An often mistake seen during this step is that folders are created leading to errors preventing the build of the module.

If folders are created, the application will not work. Be careful to always create these modules using the New, {specific} module wizard. After adding the newly created module, right-click it and build the **ui** module as shown in Figures 4-29 and 4-30.

Figure 4-29. *The newly added HTML5 module*

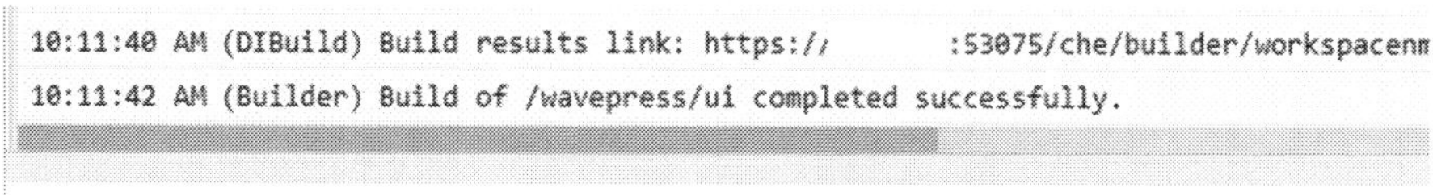

```
10:11:40 AM (DIBuild) Build results link: https://      :53075/che/builder/workspacenm
10:11:42 AM (Builder) Build of /wavepress/ui completed successfully.
```

Figure 4-30. *Console output after building the HTML5 module*

After it has successfully built, right-click it again to run it as a web application as shown in Figure 4-31.

Figure 4-31. *Running the HTML5 module as a web application*

Click the host:51027 URL displayed at the top of the console output window. The empty SAPUI5 application will launch on the next tab. This SAPUI5 application uses the default SAPUI5 theme and default version.

At this point, validate that the ui module runs and nothing else as shown in Figure 4-32.

Figure 4-32. *Default UI code in the browser after running the*
SAPUI5 HTML5 module

This is not the API yet. Since the UI is the entry point toward the API,
the application settings will need to be updated to link these two modules
together. Open the mta yaml file to see how it was updated.

Three modules should display in the graphical editor of the mta yaml
file as these are the modules created so far as displayed in Figure 4-33.

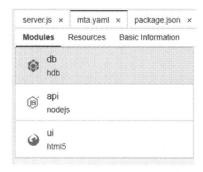

Figure 4-33. *mta file displaying modules inside the project*

In order to clear out the api error from the browser, add the wavepress
module dependencies (in the same order that they were created) following
the next steps and as displayed in Figure 4-34:

1) The **db** module (the properties in the db module should already exist correctly)

   a. Requires hdi_db container

   b. Requires the cross-schema service

Figure 4-34. *db module dependencies seen from the mta yaml file*

2) The **api** module (Figure 4-35)

   a. Requires the db and hdi_db container so that it can interact with them

   b. Requires an instance of the UAA service to authenticate the (HTTP) requests

   c. Provides a resource (to other modules) to be called externally (e.g., from the ui)

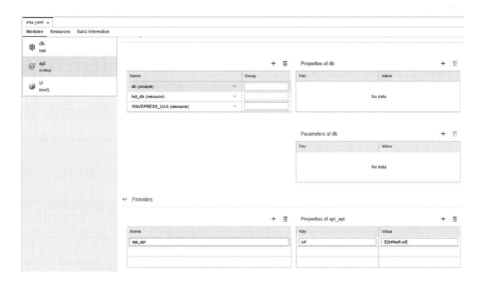

Figure 4-35. *api module dependencies seen from the mta yaml file*

3)   The **ui** module (Figure 4-36)

   a.   Requires the UAA service to authenticate the API requests

   b.   Requires the api resource to forward requests and tokens

Figure 4-36. *ui module dependencies seen from the mta yaml file*

Before building the modules again, verify that the XSA cockpit provided services to ensure the creation of the new UAA service instance.

Before (as shown in Figure 4-37):

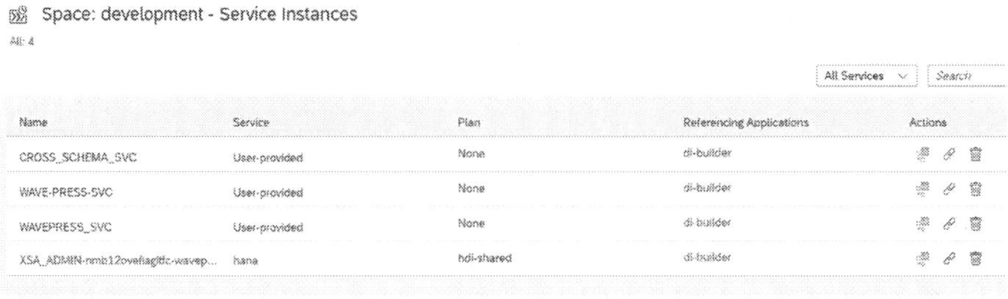

Space: development - Service Instances
All: 4

Name	Service	Plan	Referencing Applications	Actions
CROSS_SCHEMA_SVC	User-provided	None	di-builder	
WAVE-PRESS-SVC	User-provided	None	di-builder	
WAVEPRESS_SVC	User-provided	None	di-builder	
XSA_ADMIN-nmb12oveflagtfc-wavep...	hana	hdi-shared	di-builder	

Figure 4-37. *Service instances shown from the XSA cockpit*

After building the new **ui** module (and rebuilding the api module) and Figure 4-38.

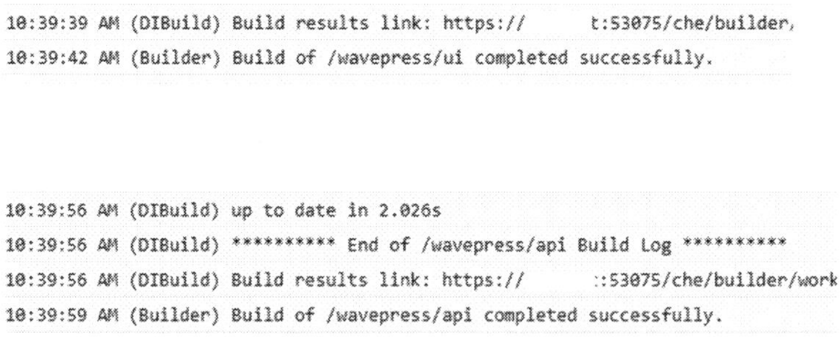

```
10:39:39 AM (DIBuild) Build results link: https://      t:53075/che/builder,
10:39:42 AM (Builder) Build of /wavepress/ui completed successfully.



10:39:56 AM (DIBuild) up to date in 2.026s
10:39:56 AM (DIBuild) ********** End of /wavepress/api Build Log **********
10:39:56 AM (DIBuild) Build results link: https://      ::53075/che/builder/work
10:39:59 AM (Builder) Build of /wavepress/api completed successfully.
```

Figure 4-38. *Console output shown after rebuilding the different modules in the project*

The newly created service instance of the XS UAA service appears in the XSA cockpit as shown in Figure 4-39.

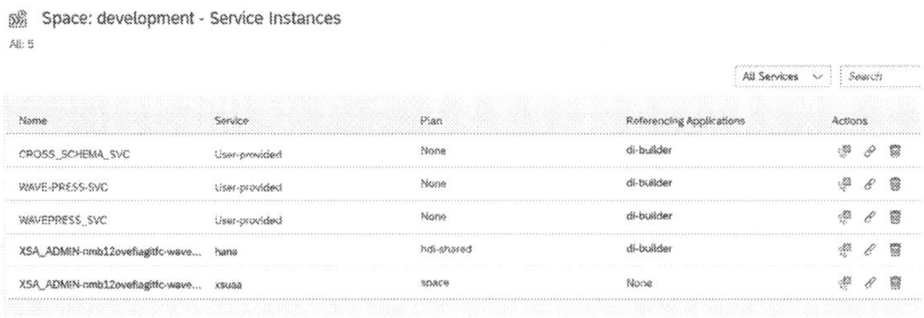

Space: development - Service Instances
All: 5

Name	Service	Plan	Referencing Applications	Actions
CROSS_SCHEMA_SVC	User-provided	None	di-builder	
WAVE-PRESS-SVC	User-provided	None	di-builder	
WAVEPRESS_SVC	User-provided	None	di-builder	
XSA_ADMIN-nmb12oveflagtfc-wave...	hana	hdi-shared	di-builder	
XSA_ADMIN-nmb12oveflagtfc-wave...	xsuaa	space	None	

Figure 4-39. *Service instances running after rebuilding the modules in the project*

148

An instance of the UAA service has been created (Figure 4-40), and it will be used by the modules to establish their relationship and dependencies. No routes have been mapped to allow access from the UI to the wavepress REST API yet. As the application will eventually require UAA authentication, the back-end module will not be directly accessible. This is a normal and secure way to disable direct access to back-end services from API consumers in software systems. Instead, the application needs to expose the ui module. Using the ui module's reroute configuration and route authentication flow, incoming requests from the browser will be passed from the ui module reaching the REST API. These are Cloud Foundry principles, not SAP HANA XSA.

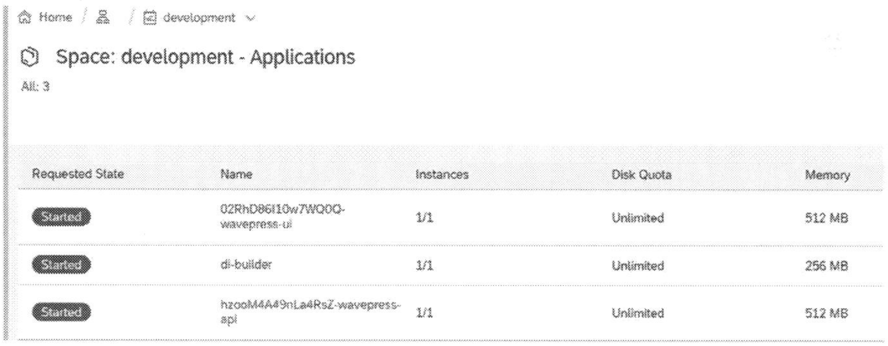

Figure 4-40. *Applications within the development space running*

The next step is to specify the routes and the authentication method in the xs-app.json file within the ui module. By default, this file looks like Figure 4-41.

Figure 4-41. *xs-app.json file configuration settings for the REST API*

Edit this file and set the ***Authentication Method*** property to route. Add the ***routes*** that will be used by the API (Figure 4-42).

1) The first route is to navigate to the Node JS REST API (following the path */wpsvc*).

2) The second route is to navigate to the XSJS backward compatibility mode service (the first service built during the initial build of the api module).

3) The third route is the "Catch all route" that reroutes the ui from any unspecified route to the webapp/ index.html file, also known as the welcomeFile.

```
xs-app.json  ×
 1 ▾  {
 2        "welcomeFile": "webapp/index.html",
 3        "authenticationMethod": "route",
 4 ▾      "routes": [
 5 ▾          {
 6                "source": "/wpsvc(.*)",
 7                "destination": "core-backend",
 8                "csrfProtection": true,
 9                "authenticationType": "xsuaa"
10 ▾          }, {
11                "source": "(.*)(.xsjs)",
12                "destination": "core-backend",
13                "csrfProtection": true,
14                "authenticationType": "xsuaa"
15 ▾          }, {
16                "source": "/(.*)",
17                "localDir" : "resources",
18                "authenticationType": "xsuaa",
19 ▾            "replace": {
20                    "pathSuffixes" : ["index.html"],
21                    "vars": ["ui5liburl"]
22                }
23        }
24        ]
25  }
```

Figure 4-42. *xs-json file including the routes and authentication method*

Continue adding routes if you want to handle different API endpoints. Routes are matched from more specific to less specific. The route settings on source/destination may use regular expressions in case you need to handle various similar routes.

At this point in the project, there is a lot of information that needs to be carefully analyzed and understood:

1) A route is a URL (it may include a port number) and the relative path it goes to (also known as root and endpoint).

2) Each of the routes contains

   a. **source** – How it arrives into the system

   b. **destination** – Using the mta, the targeted reroute

    c. **csrfProtection** – Cross-site request forgery prevention

    d. **authenticationType** – Using the xsuaa instance that was created and as it is defined in the yaml file

After the dependency connections were added, rebuild and rerun the api and the ui modules. If the build is successful as shown in Figure 4-43, navigate from the root URL where the ui module is hosted and change the relative path to the (api endpoint) default api path identified as */wpsvc*.

Run the REST API to validate that it still returns a "Hello wavepress!" message from the Node JS api module as shown in Figure 4-44. If it does, it means that the UAA service instance is being used within the application, and it is used by the application. A prompt for credentials may come up if the user has not been authenticated in the system yet. If that occurs, provide the credentials to the system.

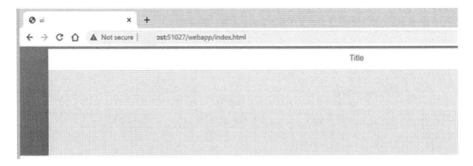

Figure 4-43. *Rerunning the application as a web application after including the route authentication*

Notice the same port used in the ui module in Figure 4-43, and the api module route (relative path) is appended after the host or domain in Figure 4-44. The browser output is generated from the nodejs service.

Figure 4-44. *Hello wavepress endpoint returning the output into the browser after making the api call*

Keep the "Hello wavepress!" endpoint as a health check endpoint temporarily (shown at line 13 in Figure 4-45). A health check endpoint is useful when verifying that a service is up and running without having to make a request that needs to reach all the way back into a database. Sometimes this health check endpoint is also known as a heartbeat.

This was a huge step in learning requests in an XSA environment. If needed, review these steps a few times to understand the integration points.

After being able to get the first health check endpoint created and validated, continue to the rest of the api development. Often, APIs start with a GET endpoint. A GET endpoint allows an API to display some data to a request.

Continue to create an endpoint that will execute a select statement against the database (utilizing an HTTP GET request) as shown in Figure 4-45. This operation represents the first integration from the api module to a database artifact. Using the calculation view created in Chapter 2, section 2.4 (CV_DEVICES), the request will look like Figure 4-45.

```
wavepress.js  ×    SQL Console 2.sql  ×
 1    /*eslint no-console. 0, no-unused-vars: 0, no-shadow: 0, new-cap: 0*/
 2    "use strict";
 3    var express = require("express");
 4  ▸ function handleErrorResponse(err, res, response) {ⁱ⁄⁰}
10
11  ▾ module.exports = function() {
12        var router = express.Router();
13        router.get("/", function(req, res) { res.send("Hello wavepress!"); });
14
15  ▾    router.get("/devices", function(req, res) {
16            var response = { data: [], errors: [] };
17            var query = " SELECT \"ID\", \"DE_UUID\", \"TEMP_F\", \"REC_DT\" FROM \"wavepress.db.models::CV_DEVICES\"; ";
18            var client = req.db;
19
20            client.prepare(query,
21  ▾            function(err, statement) {
22  ▾                if(err) {
23                        handleErrorResponse(err, res, response);
24                        return;
25                    }
26                    statement.exec([],
27  ▾                    function(err, results){
28  ▾                        if(err) {
29                                handleErrorResponse(err, res, response);
30                                return;
31                            }
32  ▾                        else {
33                                response.data = results;
34                                res.type("application/json").status(200).send(response);
35                            }
36                        });
37                });
38        });
39        return router;
40    };
```

Figure 4-45. *wavepress.js service displays the selection from the view*

The api module must be rebuilt and rerun every time code is edited or added. Make a request to GET the /devices endpoint from the api module. If the view exists and there is data in the back-end table, the request will return some data based on the select statement provided. The first and last records of the api request are shown for simplicity purposes in Figure 4-46.

```
← → C ⌂   ⚠ Not secure ████st:51027/wpsvc/devices

{
  - data: [
    - {
        ID: 2,
        DE_UUID: "2020-ABC999-1",
        TEMP_F: "73.00",
        REC_DT: "2020-04-07 10:06:00"
      },
    + {…},
    + {…},
    + {…},
    + {…},
    + {…},
    + {…},
    + {…},
    + {…},
    + {…},
    + {…},
    + {…},
    + {…},
    + {…},
    + {…},
    + {…},
    + {…},
    + {…},
    - {
        ID: 21,
        DE_UUID: "2020-DEA333-3",
        TEMP_F: "35.63",
        REC_DT: "2020-04-07 10:14:29"
      }
    ],
    errors: [ ]
}
```

Figure 4-46. *Calculation view output after calling the api from the browser*

To show a different approach to the same endpoint in the REST API, send the request via the POSTMAN REST client tool. If everything is correct, it should display the same output.

As soon as the request is successfully validated from that tool, move on to the next scenario. A POST request will be shown to demonstrate inserting and updating a record into a database table.

Using the same URL from the browser in the POSTMAN request, the API will return an ***Unauthorized*** HTTP error response as the request was initiated from the browser session from the ui module. A prompt for credentials appeared, and the ***ui*** module forwarded the request to the api module. POSTMAN will require a slightly different approach.

155

1) To simulate the authorization, an OAuth token is
    needed:

    Open POSTMAN and set the URL to the back-end
    api (not from the browser, but from the SAP Web
    IDE console).

    The URL appears after building and running the api
    module as shown in Figure 4-47.

Figure 4-47.   *Console output from the SAP Web IDE after running the node module*

To generate a token using POSTMAN, follow the
steps described here:

1)   Ensure POSTMAN has the request URL that
will be made. Ensure that the setting for collecting
cookies is enabled (displayed as a satellite icon on
the top right of the tool) as shown in Figure 4-48.

Figure 4-48.   *POSTMAN*

2)   Open the ***Authorization*** tab and select the
Type ***OAuth 2.0*** as the required type.

3)   Click the ***Get New Access Token*** button on the
right side of the screen shown in Figure 4-49.

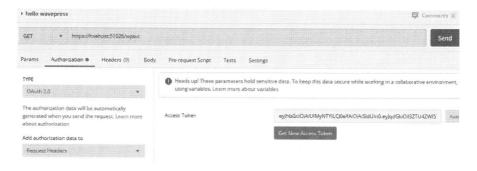

Figure 4-49. *POSTMAN Authorization to Get New Access Token*

The form for OAuth will appear. It requires it to be
populated with certain values from the environment
variables displayed from the XSA cockpit. The
correct environment variables can be collected from
the XSA cockpit by navigating to the api application
and selecting the environment variables. Look at the
values provided within the xsuaa section shown in
Figure 4-50.

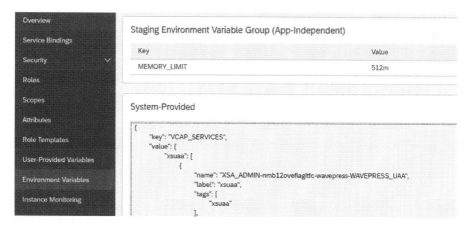

Figure 4-50. *Environment variables inside the XSA Cockpit for the
user-provided service*

Pay close attention to the VCAP_SERVICES setting, examine the xsuaa section, and obtain the following values to provide to the form from Figure 4-51:

a) The ***Token Name*** is optional. Any name may be provided.

b) The ***Access Token URL*** is the XS controller UAA service URL.

c) Append ***/oauth/token*** at the end of the URL.

d) The ***Client ID*** value shows inside the environment variable values. It is an encoded string.

e) The ***Client Secret*** is also within the same section inside the environment variable values. It is also an encoded string.

f) Click **Request Token** to generate the token that will be used in POSTMAN.

Figure 4-51. *POSTMAN Get New Access Token form*

Once requested, the token appears in the main request/response window of POSTMAN as shown in Figure 4-52.

2) Send the request and observe what type of response is received.

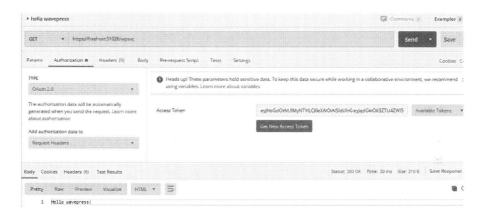

Figure 4-52. *POSTMAN Get New Access Token outputting a New Token*

After configuring the authorization token, try a POST request as shown in Figure 4-53.

To execute a POST request, additional properties are required:

1) The next endpoint to call needs to be specified as different endpoints will perform different operations in the api. This endpoint is different from the original GET operation. This time, it will serve to insert, update, or delete a record from the database.

2) Select POST as the HTTP method.

3) Provide the body of the request. The POST body requires a JSON payload (instead of executing a

GET request where the input parameters are sent as query string values in the URL).

4) The response to the POST request will also be returned as a JSON payload. The consumer of the REST API will need to know how to read that payload in order to handle additional logic.

Within the node module, and inside the wavepress.js service, create a new endpoint /device that will be used for the POST request.

```
17      router.get("/", function(req, res) { res.send("Hello wavepress!"); });
18
19 ▸    router.get("/devices", function(req, res) {⌾});
43
44 ▾    router.post("/device", function(req, res) {
45          var out = {data : req.body, errors: []};
46
47          res.type("application/json").status(200).send(out);
48      });
49
```

Figure 4-53. *wavepress API endpoints*

Send the POST request to that endpoint using a sample body payload that was retrieved from the GET request. Ensure the right structure is provided and provide the OAuth token as shown in Figure 4-54.

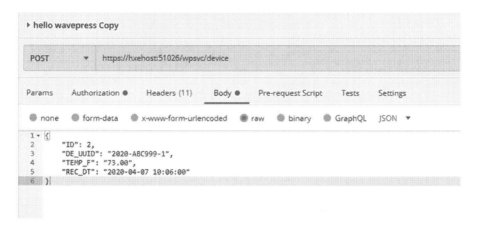

Figure 4-54. *POSTMAN POST request to update a device record*

Analyzing the request/response process and any errors will provide an understanding of the flow from any external tool to the api and from api back to the external tool (POSTMAN in this case) as a full round-trip request. The api may encounter the body being undefined, even if the body is provided in the request. It is normal to include additional logic to address other issues as they arise. The POST request's body is based on user input. All its properties and values are untrusted by the system and must be validated. If this type of issue is encountered, include a Node JS module called **body-parser** so that Express enables the interpretation of the request body coming into the API. Knowing that this module is required in the solution, it is suggested to be included as a dependency in the package.json file as shown in Figure 4-55. As a dependency, it will be downloaded, and it also must be required in the code. Open the node_ modules folder to see if it already exists as a dependency of a different module. Regardless, it is recommended to include it in the package.json file in case the original module removes its dependency in the future.

161

```
index.js  ×    wavepress.js  ×    package.json  ×
 1 ▾ {
 2 ▾    "dependencies": {
 3          "@sap/xsjs": "^5.2.0",
 4          "@sap/xsenv": "^2.0.0",
 5          "@sap/xssec" : "2.1.15",
 6          "@sap/hdbext": "6.0.0",
 7          "@sap/logging" : "4.0.2",
 8          "express" : "4.16.4",
 9          "passport" : "0.3.2",
10          "body-parser": "1.18.3"
11       },
12       "description": "wavepress REST API",
13 ▾    "devDependencies": {
14       },
15       "files": [],
16       "main": "server.js",
17       "name": "api",
18 ▾    "scripts": {
19          "start": "node server.js"
20       },
21 ▾    "engines": {
22          "node": "8.x"
23       },
24       "version": "1.0.0"
25    }
```

Figure 4-55. *package.json inside the Node.js module including the body-parse module*

After it is included, import it on the wavepress service implementation as shown in Figure 4-56.

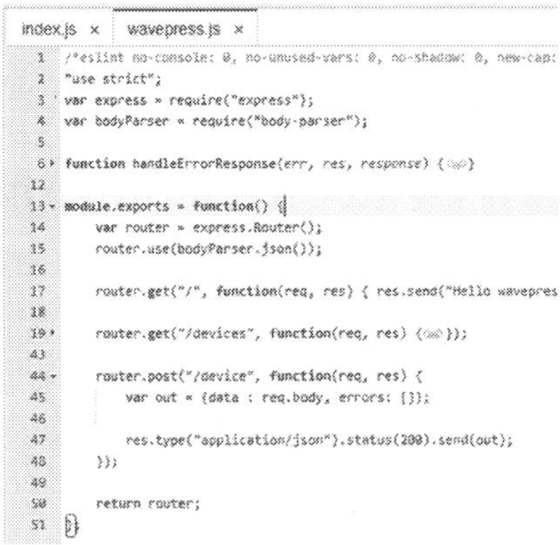

```
index.js  ×    wavepress.js  ×
 1   /*eslint no-console: 0, no-unused-vars: 0, no-shadow: 0, new-cap:
 2   "use strict";
 3   var express = require("express");
 4   var bodyParser = require("body-parser");
 5
 6 ▸ function handleErrorResponse(err, res, response) {…}
12
13 ▾ module.exports = function() {
14      var router = express.Router();
15      router.use(bodyParser.json());
16
17      router.get("/", function(req, res) { res.send("Hello wavepres
18
19 ▸    router.get("/devices", function(req, res) {…});
43
44 ▾    router.post("/device", function(req, res) {
45         var out = {data : req.body, errors: []};
46
47         res.type("application/json").status(200).send(out);
48      });
49
50      return router;
51   }
```

Figure 4-56. *wavepress service importing the service implementation (lines 4 and 15)*

After rebuilding the api module, proceed by rerunning it again. After it starts running, return to POSTMAN and send a follow-up POST request as shown in Figure 4-57.

Figure 4-57. *POSTMAN update request after including the body-parser*

This time, the request is successful as Express can read the request body that was provided. It was also returned as a response to the POSTMAN tool. The round trip on the api module has been validated.

After successfully sending a request to the POST endpoint and validating that it works, wire up the call to execute the stored procedure that was coded in Chapter 2, section 2.4. It is recommended to break down the calls to the api and the implementation from Node JS against the database in small pieces, ensuring that each piece of the integration scenario works prior to the full scenario integration. Apply the same principle moving from POST to the DELETE HTTP method implementation. Following this approach will ensure there are no issues in the default settings. Any issues that come up can be easily solved by breaking down the integration points. Divide and conquer!

The JavaScript code that will load stored procedures to update/insert a record is shown in Figure 4-58.

Line 53 gets the db client object, and this client needs to be provided when loading the stored procedure using the hdbext (sap) module. The stored procedure is loaded into the Node.js runtime in line 56 as an object, and then it is executed in line 63. Add any business logic to validate inputs or to format input parameters to the stored procedure before executing it on line 63. The first argument of the stored procedure is to provide any input parameters that it expects. In the following example, the device array is passed in. Notice the device array contains one object, and it is the request's body object.

Any output parameters will be returned from the callback provided as a second argument of the stored procedure. In the following example, the stored procedure returns an output parameter named OUT_ID.

It is very nice to see the entire handling of the request, the business logic, and also calling the SAP HANA database objects as JavaScript objects without having to do JSON parsing either going in or out of the Node.js implementation as shown in Figure 4-58.

```
50      // using post to insert/update
51 ▾    router.post("/device", function(req, res) {
52          var response = { data : [], errors: []};
53          var client = req.db;
54          var hdbext = require("@sap/hdbext");
55
56 ▾      hdbext.loadProcedure(client, null, "wavepress.db.procs::SP_DEVICE_UPSERT", function(err, sp) {
57 ▾          if(err) {
58              handleErrorResponse(err, res, response);
59              return;
60          }
61          var device = [];
62          device.push(req.body);
63 ▾        sp({IN_DEVICE: device }, function(err, parameters, result) {
64 ▾            if(err){
65                handleErrorResponse(err, res, response);
66                return;
67              }
68              response.data = parameters.OUT_ID;
69              handleCorrectResponse(res, response);
70          });
71      });
72    });
```

Figure 4-58. *Device endpoint to call a SAP HANA stored procedure*

Repeat the steps of rebuilding and rerunning the api module, enabling the latest code to be available for the api.

After the api module is running, run the POST request. Start by getting the raw JSON from an existing record (obtain it from the output of the GET request) as shown in Figure 4-59.

Figure 4-59. *POSTMAN GET request to get a list of devices*

Copy the JSON payload as the body of the next POST request on a different POSTMAN tab. Update a property in the body, namely, the TEMP_F value from 74 to 72 to test out the update operation as shown in Figure 4-60. Click **Send**.

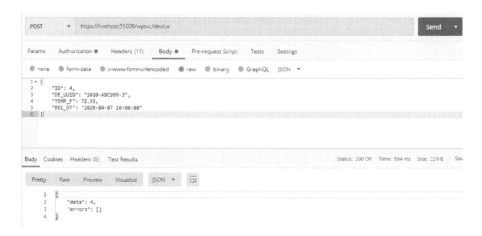

Figure 4-60. *POSTMAN POST request and response from the wavepress service after updating a record*

If the structure of the body is correct and the stored procedure executes successfully, the response of the stored procedure will return the ID value that was updated as the output of the response. The response can be validated by rerunning the GET endpoint and seeing the value updated in real time as shown in Figure 4-61.

Figure 4-61. *POSTMAN request validating the updated record*

Run an insert operation to simulate a new record getting into the database. Notice that the ID property is null and ensure the TEMP_F is of data type decimal, not a string, avoiding a data type exception during the execution of the SQL statement as shown in Figure 4-62.

Adding logic in the nodejs api to handle erroneous inputs and to handle the output response is a good practice to let the consumer of the api have any incorrect behavior or expected values.

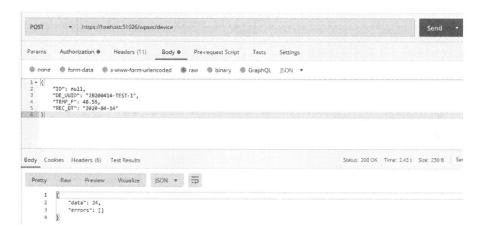

Figure 4-62. *POSTMAN request sending a null ID to insert a record*

Rerun the validation through the GET HTTP request endpoint as it was done before and shown in Figure 4-63.

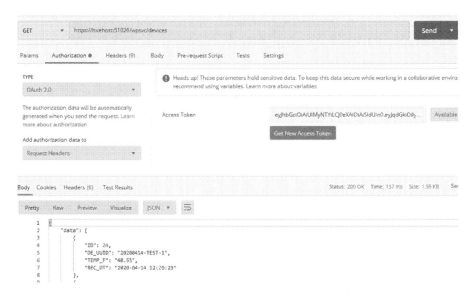

Figure 4-63. *POSTMAN request to validate the last insert via GET request*

Test the delete endpoint next as shown in Figure 4-64. Keep in mind in this endpoint, the ID of the record wished to be deleted must be provided. The ID of the record is passed as an argument of the URL rather than in the body of the request. The route of the delete endpoint looks like Figure 4-64.

This time, the query string parameter contains the ID of the record that will be deleted. The ID is passed as the input parameter, IN_DEVICE_ID, of the stored procedure.

```
74  router.delete("/device/:id*?", function(req,res) {
75      var response = { data: [], errors: []};
76      var client = req.db;
77      var hdbext = require("@sap/hdbext");
78      hdbext.loadProcedure(client, null, "wavepress.db.procs::SP_DEVICE_DELETE", function(err, sp) {
79          if(err) {
80              handleErrorResponse(err, res, response);
81              return;
82          }
83          var deviceId = req.params.id || null;
84          sp({ IN_DEVICE_ID: deviceId }, function(err, parameters, result) {
85              if(err){
86                  handleErrorResponse(err, res, response);
87                  return;
88              }
89              response.data = parameters.IS_DELETED === 1 ? "deletion successful" : "error while deleting " + (deviceId.toString() || "");
90              handleCorrectResponse(res, response);
91          });
92      });
93  });
```

Figure 4-64. *wavepress service showing endpoint for delete operation*

From POSTMAN, the DELETE request appears as displayed in Figure 4-65.

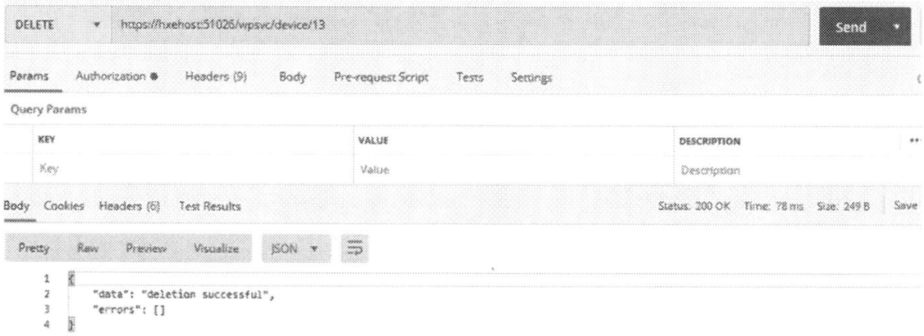

Figure 4-65. *POSTMAN request sending a delete request*

To validate the request, rerun the GET statement, ensuring that the record matching device ID = 13 no longer exists in the response. Notice the gap in the sequence of the items displayed from the SQL console in Figure 4-66.

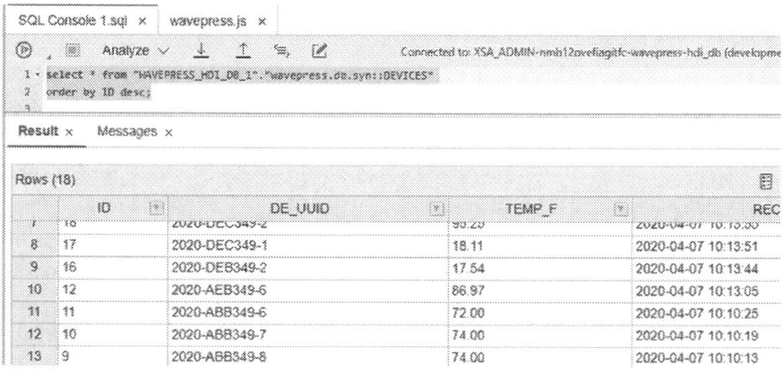

Figure 4-66. *Database Explorer SQL console validating the record does not exist*

The delete operation can be validated from POSTMAN using the GET request once again as shown in Figure 4-67.

Figure 4-67. *POSTMAN request to validate the deleted record does not return from the GET request*

Debugging the Node JS (api) module

Debugging is the process of setting breakpoints in the code (interrupting execution) in order to troubleshoot a bug or examine values at specific points in the code. Debugging is a fundamental part of any software development, and it should be done carefully and wisely.

To run a debugging scenario in the SAP Web IDE, take one of the endpoints in the **api** module, set up a few breakpoints, and start a debugging session as shown in Figure 4-68. The current subsection shows how to set up debugging sessions.

In the **api** module, open the wavepress service and do the following:

1) On the code lines (left perimeter of the screenshot), click one of the lines to set a breakpoint.

2) On the right navigation panel, select the fifth icon from the top to open the debugging pane. Notice at the bottom of the debugging pane, it displays the two lines where debugging points were set.

3) Near the top of the debugging pane, the Active Session label is seen, and a drop-down control shows the selected active session. Next to the drop-down control, there are icons that establish a session, detach a session, and show information about a session.

Figure 4-68. *Web IDE debugging pane*

4)   Click the attach debugger icon (Figure 4-69).

Figure 4-69. *Web IDE debugging points*

5)   Select the api session (by selecting run script start) and click the OK button (Figure 4-70).

Figure 4-70. *Web IDE attaching a debugging to target*

Notice the breakpoints have a checkmark and the debugger pane shows an active session. If the debugger is active, proceed to run a request from POSTMAN as shown in Figure 4-71.

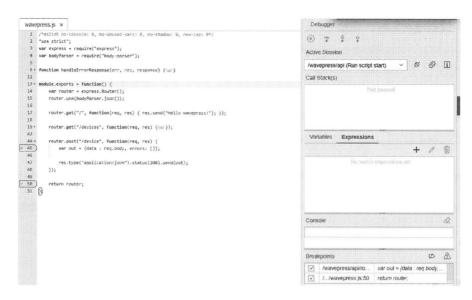

Figure 4-71. *Web IDE debugger in action*

Firing the request from POSTMAN executes the code in the **api**, and it stops at the first breakpoint that was set, as shown in Figure 4-72.

Figure 4-72. *Web IDE and debugging pane during an active session*

As the debugger has stopped where it was expected, examine the call stack on the right. The call stack shows the statements executed to stop at the breakpoint.

173

Navigation within the debugger can be performed from the top icons (under the Debugger section) or by using some function keys in the keyboard. It is very simple, and it is done in the same way as it is done in other development tools.

The **play button** will advance the code execution from the current line and will only stop again if there is another breakpoint in front of the current line of execution. If no other breakpoint is set, then the execution will continue until the execution ends. This functionality can also be executed by pressing F8 on the keyboard.

The next icon (arrow pointing right) is to **Step over**. The step over feature allows developers to continue execution and allows execution to jump over code inside nested functions. Step over can be achieved by pressing F10 on the keyboard.

Next, there is the **Step into** feature. This feature allows execution to go to the next line, and it gets into nested functions as well. Use this feature if you want to examine a line-by-line approach. The same can be achieved by pressing the F11 key on the keyboard.

Finally, within the debugging subsection, there is **Step out** (of the current function). Use this feature to step out of a function when nested inside. The same feature can be achieved by pressing Shift+F11 on the keyboard.

The **Variables** tab (Figure 4-73) is used for local variable values, global variable values, and any other type of scope variable from the current code execution, including JS closures.

Figure 4-73. *Debugging variables*

In addition to examining the current variable values, *Expressions* can also be set (Figure 4-74). Click the *Expressions* tab and click the plus sign under **Expressions** to add an expression.

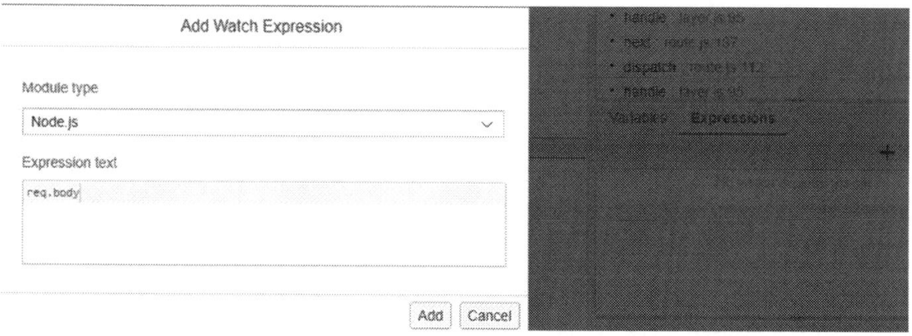

Figure 4-74. *Debugging watch expression*

If set correctly, the Expressions tab will show the expression added. Any expression that are objects will allow further expansion to see its nested properties as shown in Figure 4-75.

Figure 4-75. *Debugging expressions*

Under the Expressions section, the *Console* section can be selected (Figure 4-76). This section allows entering variable names to see values as quick expression evaluations, while the Expressions tab will continue to display the current value even after advancing code execution.

Figure 4-76. *Debugging console showing variable values*

Another scenario that must be described is debugging stored procedures from the SAP Web IDE. Stored procedures are database artifacts that act as functions, and they perform read, create, update, and delete operations against a database table. Stored procedures accept input and output parameters of simple (NVARCHAR, INT, DECIMAL, SECCONDDATE) or complex types (structures as tables). In Chapter 2, section 2.4, the SP_DEVICE_UPSERT stored procedure was created to be used from the scenario describing the /device endpoint created in Chapter 4, section 4.3. When developing APIs, sometimes the issue in context may be external to the JavaScript code; however, if the developer can troubleshoot SqlScript in the SAP HANA database, that results in an end to end development scenario.

Begin by opening the Database Explorer and select the stored procedure. Right-click it and select Open for Debugging as shown in Figure 4-77.

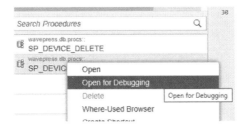

Figure 4-77. *Stored procedure debugging*

After clicking the option "Open for Debugging," there will be a pop-up showing the different ways to attach the debugger. Leave the "SQL console connections" selected as the example will show the debugging of this stored procedure from the Database Explorer. Select OK in Figure 4-78.

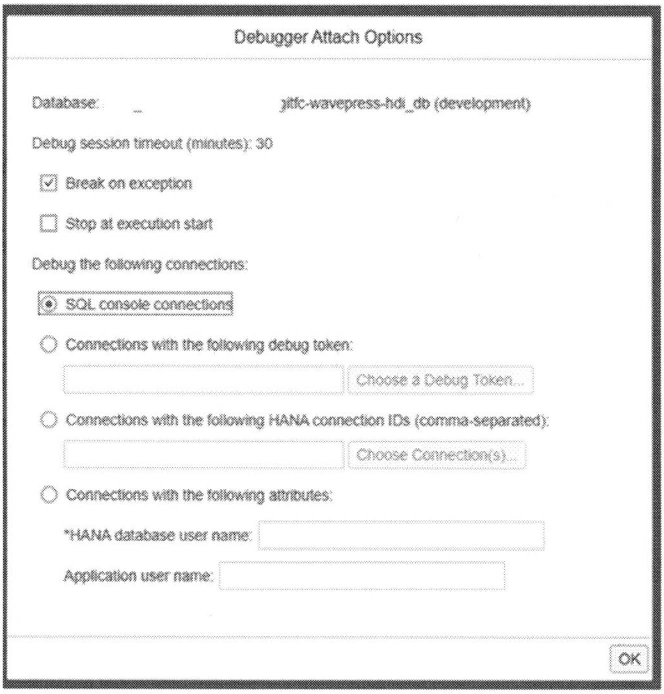

Figure 4-78. *Attaching a debugger to a stored procedure via the SQL console connection*

The stored procedure will be opened in the Database Explorer, and debugging points can be set in a similar way they were set in the Node JS API. Add some debugging points on the left perimeter of the screen where the line numbers are shown. See Figure 4-79.

177

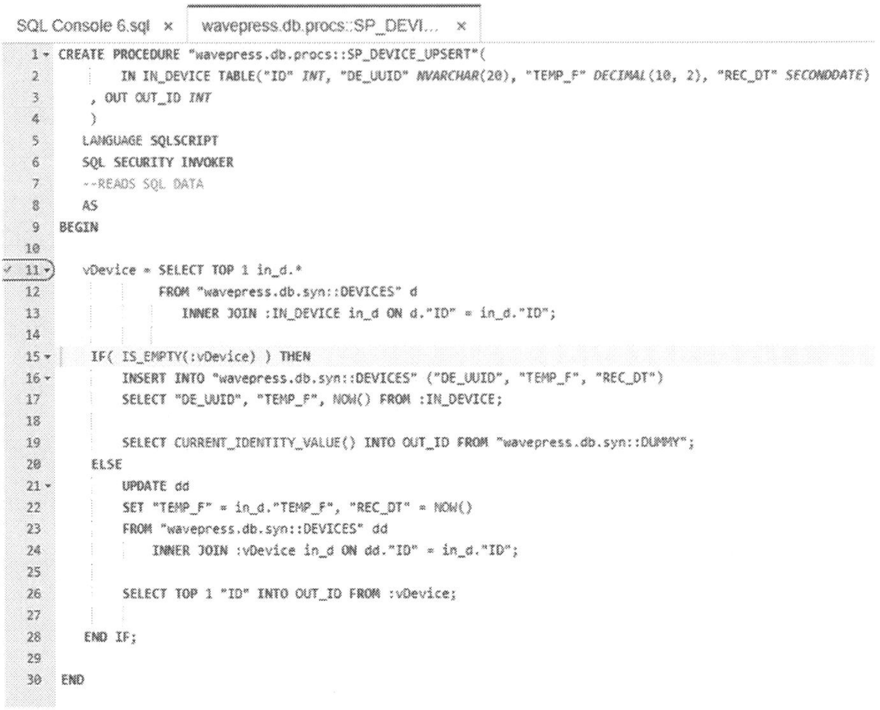

```
SQL Console 6.sql  ×    wavepress.db.procs::SP_DEVI...  ×

 1 ▾ CREATE PROCEDURE "wavepress.db.procs::SP_DEVICE_UPSERT"(
 2        IN IN_DEVICE TABLE("ID" INT, "DE_UUID" NVARCHAR(20), "TEMP_F" DECIMAL(10, 2), "REC_DT" SECONDDATE)
 3      , OUT OUT_ID INT
 4      )
 5      LANGUAGE SQLSCRIPT
 6      SQL SECURITY INVOKER
 7      --READS SQL DATA
 8      AS
 9   BEGIN
10
11 ▾     vDevice = SELECT TOP 1 in_d.*
12              FROM "wavepress.db.syn::DEVICES" d
13                  INNER JOIN :IN_DEVICE in_d ON d."ID" = in_d."ID";
14
15 ▾     IF( IS_EMPTY(:vDevice) ) THEN
16 ▾         INSERT INTO "wavepress.db.syn::DEVICES" ("DE_UUID", "TEMP_F", "REC_DT")
17          SELECT "DE_UUID", "TEMP_F", NOW() FROM :IN_DEVICE;
18
19          SELECT CURRENT_IDENTITY_VALUE() INTO OUT_ID FROM "wavepress.db.syn::DUMMY";
20      ELSE
21 ▾         UPDATE dd
22          SET "TEMP_F" = in_d."TEMP_F", "REC_DT" = NOW()
23          FROM "wavepress.db.syn::DEVICES" dd
24              INNER JOIN :vDevice in_d ON dd."ID" = in_d."ID";
25
26          SELECT TOP 1 "ID" INTO OUT_ID FROM :vDevice;
27
28      END IF;
29
30   END
```

Figure 4-79. *Debugging stored procedure*

On the right side of the SAP Web IDE, observe the debugging pane as shown in Figure 4-80.

Figure 4-80. *Debugger during stored procedure debugging session*

Open a SQL console window and prepare the call to the stored procedure. Since the SP_DEVICE_UPSERT stored procedure uses a table type as an input parameter, the example shows a temporary table with the same structure as the input parameter. A record added to the temporary table simulates the same structure. The select statement before the call to the stored procedure is to validate the data is present.

Highlight the CALL "WAVEPRESS_HDI_DB_1". "wavepress. db.procs::SP_DEVICE_UPSERT"(....) statement and click the run icon (green play button on the top left of Figure 4-81).

```
SQL Console 6.sql   ×      wavepress.db.procs::SP_DEVI...   ×

 ⊳      ▦    Analyze ∨   ↓   ↑   ⇆   ✎        Connected to: XSA_ADMIN-nmb12ovefiagitfc-wavepress-hdi_db (development

 1
 2   create local temporary table #a ("ID" INT, "DE_UUID" NVARCHAR(20), "TEMP_F" DECIMAL(10, 2), "REC_DT" SECONDDATE);
 3
 4 ⁃ insert into #a(ID, DE_UUID, TEMP_F, REC_DT)
 5   values(25,'debugTest2', 100, now())
 6
 7   select * from #a
 8
 9
10   CALL "WAVEPRESS_HDI_DB_1"."wavepress.db.procs::SP_DEVICE_UPSERT"(IN_DEVICE => #a ,OUT_ID => ?);
```

Figure 4-81. *SQL console to start debugging a stored procedure that takes a table type as input parameter*

If the input parameter has the right structure, the debugging session is started, and the execution of the stored procedure stops at the breakpoint as shown in Figure 4-82.

```
SQL Console 6.sql   ×      wavepress.db.procs::SP_DEVI...   ×

 1 ⁃  CREATE PROCEDURE "wavepress.db.procs::SP_DEVICE_UPSERT"(
 2           IN IN_DEVICE TABLE("ID" INT, "DE_UUID" NVARCHAR(20), "TEMP_F" DECIMAL(10, 2), "REC_I
 3         , OUT OUT_ID INT
 4         )
 5       LANGUAGE SQLSCRIPT
 6       SQL SECURITY INVOKER
 7       --READS SQL DATA
 8       AS
 9   BEGIN
10
11 ⁃      vDevice = SELECT TOP 1 in_d.*
12              FROM "wavepress.db.syn::DEVICES" d
13                  INNER JOIN :IN_DEVICE in_d ON d."ID" = in_d."ID";
14
15 ⁃      IF( IS_EMPTY(:vDevice) ) THEN
16 ⁃          INSERT INTO "wavepress.db.syn::DEVICES" ("DE_UUID", "TEMP_F", "REC_DT")
17           SELECT "DE_UUID", "TEMP_F", NOW() FROM :IN_DEVICE;
18
19           SELECT CURRENT_IDENTITY_VALUE() INTO OUT_ID FROM "wavepress.db.syn::DUMMY";
20       ELSE
21 ⁃          UPDATE dd
22           SET "TEMP_F" = in_d."TEMP_F", "REC_DT" = NOW()
23           FROM "wavepress.db.syn::DEVICES" dd
24               INNER JOIN :vDevice in_d ON dd."ID" = in_d."ID";
25
```

Figure 4-82. *Debugging stopped at the debugging point in stored procedure*

In the debugger (Figure 4-83), the call stack shows where the current execution is. Furthermore, the context Variables and Expressions tabs show the available information from the debugging session for analysis. These same tabs and features were showcased in this chapter.

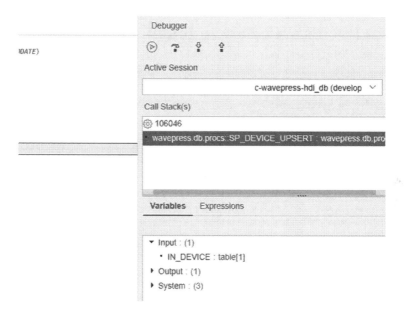

Figure 4-83. *Debugging call stack and variables while debugging stored procedure*

Continue moving execution with the Step over, Step into, or Resume execution features of the debugger. Once debugging is complete in this object, then the output (if any) will be displayed back in the SQL console where the debugging session started.

As shown in this chapter, typically debugging features are used in conjunction with each other such as using debugging variables or writing expressions outputting to the console and using special keyboard keys to move from one statement of code to another (or from one debugging point to another) while debugging Node JS or XSJS code from the SAP Web IDE. The debugging controls are used in any typical scenario of API software development in the SAP HANA XSA environment. Remember that debugging occurs on the server side, and the external tools facilitate the launch of these debugging sessions (for POST requests using POSTMAN or the browser window for GET requests). Using these features often allows developers to understand the flow and execution logic of a software program as well as assists in understanding the behavior of the SAP Web IDE and the Database Explorer. Whether the situation calls for debugging JavaScript or SqlScript code, this section has shown both situations, how to prepare to handle them and the suggested approach to execute your development and debugging.

Conclusion

The chapter concludes with showing all the tricks and tips that any experienced developer would expect as far as being able to navigate the tools in the development cycle, being able to set up debugging points, and being able to debug every aspect of a software system, whether it is in the back end on a stored procedure or in the microservice providing a robust api; this chapter walks along each of the integration points and identifies the various ways on how to be successful in completing an end-to-end development scenario which includes the GET, POST, and DELETE HTTP methods.

CHAPTER 5

Deployment scenarios of HANA XSA

A software system is not completed until it is delivered in a production environment. The book will showcase and explain how to assign version control to the microservice being developed in the exercises shared in this book. This chapter will also explain how to perform deployments of software in the environment and also how to deploy to the SAP Cloud Platform.

MTA project and versioning

An important stage of software development is versioning and release cycle. During the different stages of a software product life span, development teams will assign versioning to their product to comply with auditing and documentation. In SAP HANA 1.x, there was no easy way to keep product versioning; however, there are features in the SAP HANA XSA architecture that facilitate the version control of software.

© Sergio Guerrero 2020
S. Guerrero, *Microservices in SAP HANA XSA*, https://doi.org/10.1007/978-1-4842-6118-7_5

In SAP HANA XSA, and as it has been demonstrated in previous chapters, the mta yaml file contains metadata properties relating to the multitarget application. This file also contains module dependencies, and it is used by the application builder to configure relationships between modules. An additional feature of this file is that it can keep the version of the build as a runtime property.

After the latest application build, the mta yaml file looks like Figure 5-1.

```
mta.yaml  ×
 1   ID: wavepress
 2   _schema-version: '2.1'
 3   version: 0.0.1
 4 ▾ modules:
 5 ▾   - name: db
 6       type: hdb
 7       path: db
 8 ▾     requires:
 9 ▾       - name: hdi_db
10 ▾         properties:
11             TARGET_CONTAINER: '~{hdi-container-name}'
12 ▾       - name: WAVEPRESS_SVC
13           group: SERVICE_REPLACEMENTS
14 ▾         properties:
15             key: ServiceName_1
16             service: ~{wavepress-service-name}
17 ▾   - name: api
18       type: nodejs
19       path: api
20 ▾     provides:
21 ▾       - name: api_api
22 ▾         properties:
23             url: ${default-url}
24             service: '~{wavepress-service-name}'
25 ▾     requires:
26         - name: db
27         - name: hdi_db
28         - name: WAVEPRESS_UAA
29 ▾   - name: ui
30       type: html5
31       path: ui
32 ▾     requires:
33         - name: WAVEPRESS_UAA
34 ▾       - name: api_api
```

***Figure 5-1.** mta yaml*

It is evident that the property named ***version*** is indeed the one offering the feature within the mta yaml artifact. So far, the provided examples have instructed us to build and run modules to see the REST API in action.

The next build demonstration will be at the application level (in contrast to the module-level builds).

Select the project, right-click it, and select Build as shown in Figure 5-2.

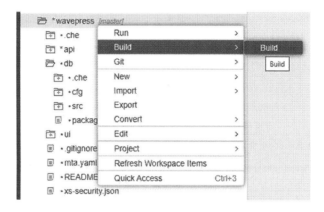

Figure 5-2. *Application build*

This step will take longer to run than building the modules independently due to the application builder gathering information for each of the modules and running a collective build. Look at the bottom right side of the SAP Web IDE. It shows the application build is happening. When building at the application level completes, open the console, and notice the log displaying the complete successful message as displayed in Figure 5-3.

```
7:58:30 AM (DIBuild) [INFO] Creating MTA archive
7:59:42 AM (DIBuild) [INFO] Saving MTA archive wavepress_0.0.1.mtar
7:59:44 AM (DIBuild) [INFO] Done
7:59:44 AM (DIBuild) ********** End of /wavepress Build Log **********
7:59:44 AM (DIBuild) Build results link: https://    ost:53075/che/builder/workspacenmb12ovefiagitfc/download-all/8d2d655f-5b70-4fea-ba
7:59:44 AM (Builder) The .mtar file build artifact was generated under /mta_archives/wavepress.
7:59:46 AM (Builder) Build of /wavepress completed successfully.
```

Figure 5-3. *Application build console output*

After validating the mta build from the console, a new folder will be shown within the workspace hierarchy, as shown in Figure 5-4, called ***mta_archives***. This folder contains subfolders that will match applications built within the workspace. Each folder will contain the archive files generated from each application-level build.

Figure 5-4. *mta archives under workspace*

The multitarget application archive generated is known as an ***mtar*** file. The archive file is generated with the version specified in the mta yaml file as part of its name such as that of Figure 5-5. Ensure the version in the mta file and the version within the name of the mtar match. If the application is rebuilt, the mtar file is overwritten by the new build. Beware of the process when rebuilding the application to avoid an overwrite in your application build.

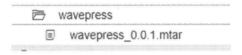

Figure 5-5. *mtar generated file*

If a different version of the application is needed, update the mta yaml file, increasing the version number from 0.0.1 to 0.0.2 depending on the desired version as shown in Figure 5-6.

Best practices in software versioning describe the version pattern as X.Y.Z, where

1) **X** represents a major release number. A major release means that various features were added to the software and released.

2) **Y** represents a minor release number. A minor release means that a small feature was added to the software and released.

3) **Z** represents a patch number. A patch number means that a different build of the software was added and released.

In the application, open the mta yaml file and increase the version number to showcase the versioning process as displayed in Figure 5-6.

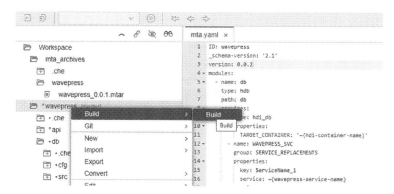

Figure 5-6. *mta yaml version update*

Select the application, right-click it, and rebuild it once more as shown in Figure 5-7.

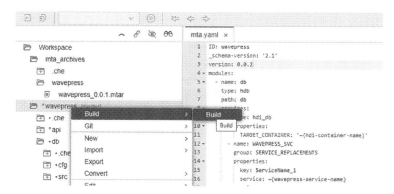

Figure 5-7. *Rebuilding application*

Return to the mtar build folder and notice the newly generated mta archive file. It matches the updated version from the mta yaml file. Notice within the same folder that both the initial 0.0.1 and 0.0.2 archives are kept. The same will hold true if the version property continues to be updated again within the mta yaml file. This time the version will be updated to 0.0.3 and so on as needed as shown in Figure 5-8.

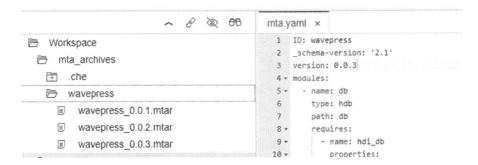

Figure 5-8. *mta yaml versions and generated mtars*

SAP Web IDE deployments

Deployments from the SAP Web IDE can be done by selecting the mtar file from the available set of mtars and by selecting the menu item **Deploy** from the SAP Web IDE top menu. This section of the chapter is unrelated to development tasks; however, it is important to understand how to perform deployments in an XSA environment. Both developers and XSA cockpit admins may be involved in the series of tasks described going forward.

There are two types of deployment available:

1) SAP HANA XS advanced (showcased here)

   This option is used if on-premise deployment is performed. Knowing the (CF architecture) Organization and Space is required to deploy an application.

In the SAP Web IDE, select the mtar file. Follow by
selecting the "Deploy" menu item from the navigation.
Select "Deploy to XS Advanced" as shown in Figure 5-9.

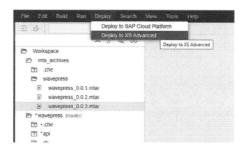

Figure 5-9. *mtar deploy to XSA*

After selecting the XS Advanced option from the Deploy
menu, follow the prompt to specify the Organization and
Space for the deployment as shown in Figure 5-10. The
same application could be deployed to different Spaces
if one Space is needed for development and another
Space may be needed for testing purposes. Deploying to
different Spaces will eventually generate a different URL:
PORT where the application is deployed to.

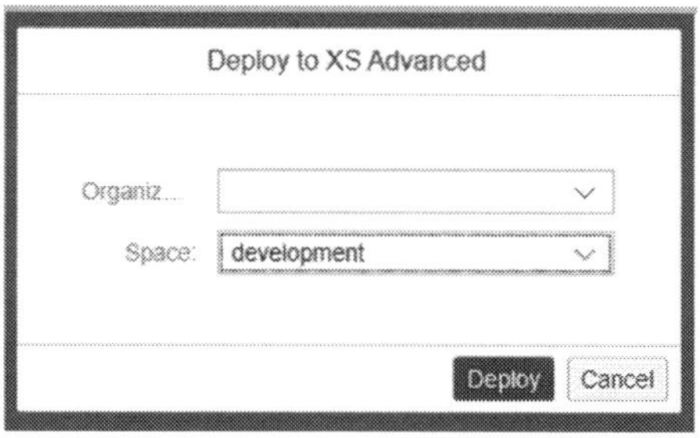

Figure 5-10. *mtar deploy to Organization/Space*

Click the "Deploy" button.

The deployment of the application may take a few minutes as the SAP Web IDE is using the mtar (binary file generated) to deploy the application to the on-premise XSA environment. Continue to monitor the console log for any issues or successful deployment messages as displayed in Figure 5-11.

Figure 5-11. *mtar deployment console message*

The ui application is running on the host:51053 port which means that the deployed application will be launched from a different port than the same application that was used during the development exercise. Along the same lines, it is expected that developers/admins will be able to monitor these applications from the XSA Advanced Cockpit as shown in Figure 5-12.

Navigate to the XSA cockpit, into the Organization and
Space, and notice the deployed applications.

◎ Space: development - Applications

All: 6

Requested State	Name	Instances	Disk Quota	Memory
Started	02RhD86I10w7WQ0Q-wavepress-ui	1/1	Unlimited	512 MB
Started	api	1/1	Unlimited	1024 MB
Stopped	db	0/1	Unlimited	256 MB
Started	di-builder	1/1	Unlimited	256 MB
Started	hzooM4A49nLa4RsZ-wavepress-api	1/1	Unlimited	512 MB
Started	ui	1/1	Unlimited	1024 MB

Figure 5-12. *XSA applications running in development Space*

Running the application from the newly bound port
should return the output of the API.

Validate this deployment step by navigating to the URL
and PORT where the ui module is running, and it will
display the ui module as shown in Figure 5-13.

Figure 5-13. *(Deployed) applications running from different ports*

Similarly, validate the api module by changing the relative path to navigate to the /wpsvc url (displaying the "Hello wavepress!" message on the browser) as it was shown before and also in Figure 5-14.

Figure 5-14. *Browser output from a deployed application*

Finally, validate the devices endpoint to retrieve the list of devices as shown in Figure 5-15 (which executes the CV_DEVICES view).

Figure 5-15. *Browser output for /devices endpoint*

The other endpoints to insert, update, and delete devices can also be validated. Since this section follows the same steps as Chapter 4, validate the steps in Chapter 4 to validate the other HTTP methods simply by updating the relative path and HTTP method and send the request from POSTMAN.

The deployment process from the SAP Web IDE to XSA environment on-premise has been successfully validated now.

2) Deploying to the SAP Cloud Platform Cloud Foundry environment

When using this option, an active account is required in the SAP Cloud Platform (or a free trial account within the SAP Cloud Platform) in order to deploy an MTA application.

Furthermore, as described in Chapter 1 within the CF subsection, it will be a prerequisite to already have an Organization and a Space where the application will be deployed. If this is not the case, proceed to create the Organization and Space for this exercise.

Log in to a SAP Cloud account (or trial account) as displayed in Figure 5-16. Make sure to consult the types of SAP Cloud licenses available if you need to run applications in a production environment.

Figure 5-16. *SAP Cloud Platform*

Ensure that CF is enabled (or enable it after clicking "Enter Your Trial Account") as shown in Figure 5-17.

Figure 5-17. *Enable CF in SAP Cloud*

After the CF environment is enabled, it will look like Figure 5-18. Beware that the CF environment is enabled for 30 days initially. Should the environment need to be enabled for longer than 30 days, making a request to extend the trial period is required to prevent it from being deactivated and possibly losing any changes.

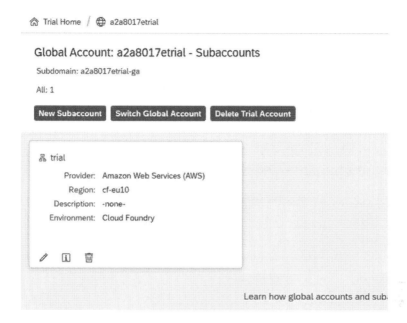

Figure 5-18. *SAP Cloud Platform – CF*

Click the tile in the content section.

The next screen displays the CF settings (API Endpoint, Organization, and Space) that are needed to deploy the application from the SAP Web IDE. Currently, there is no application or service deployed to it as seen in Figure 5-19.

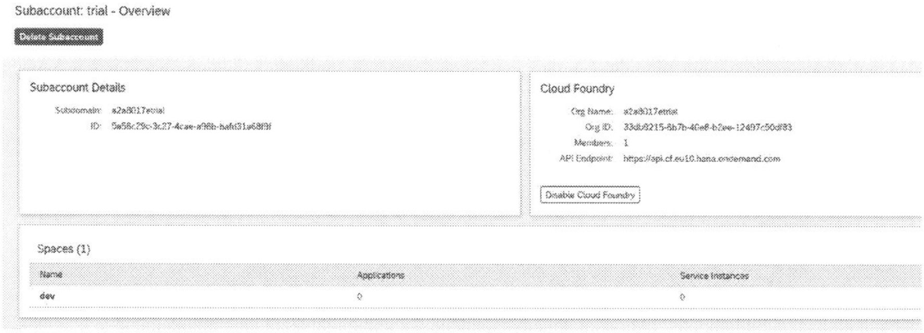

Figure 5-19. *SAP Cloud Platform CF Organization/endpoint*

Although the REST API was deployed and tested in the on-premise scenario, the same deployment configuration of the MTA yaml file may not be correct when deploying to a CF environment. Be aware that in a CF environment, services and applications share resources. During the build of this API solution, additional settings were added to the mta yaml file to comply with the CF deployment.

The settings added are as follows.

Within the hdi-db resource, add the key-value pairs inside the parameters section:

a)   service : hanatrial

b)   service-plan : hdi-shared

In the SAP Web IDE, select one of the mtar files, right-click, and select Deploy to SAP Cloud Platform as shown in Figure 5-20.

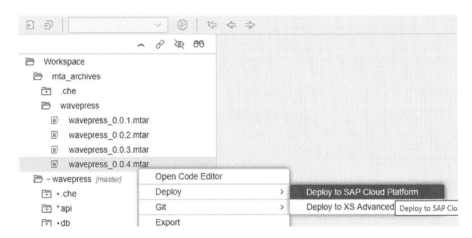

Figure 5-20. *Deploying to SAP Cloud Platform*

The SAP Web IDE has an interface that requires the CF API Endpoint details, and it is shown in Figure 5-21.

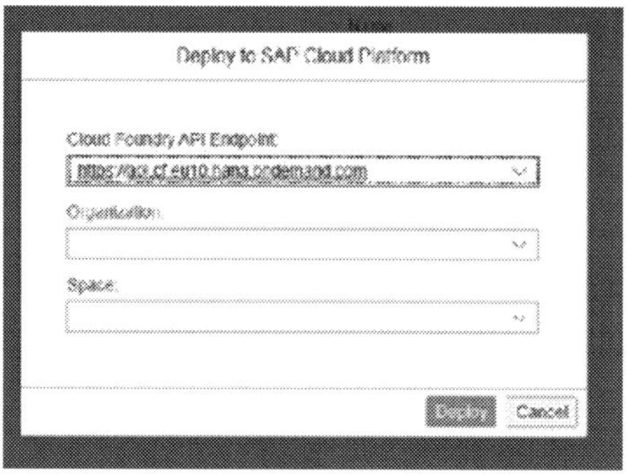

Figure 5-21. *SAP Cloud endpoint, Organization, Space*

After the API endpoint is provided, the SAP Web IDE will prompt for credentials to the CF account as displayed in Figure 5-22.

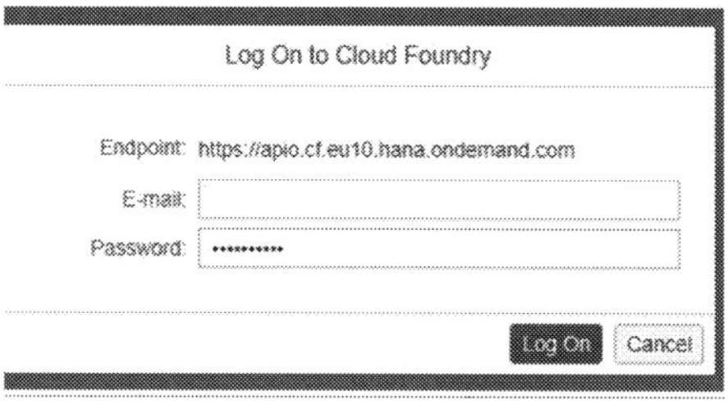

Figure 5-22. *Credentials for the CF account*

If the CF account is authenticated with the credentials provided against the CF Endpoint, the prompt will display the Organization and Space details after being retrieved from the CF account as shown in Figure 5-23.

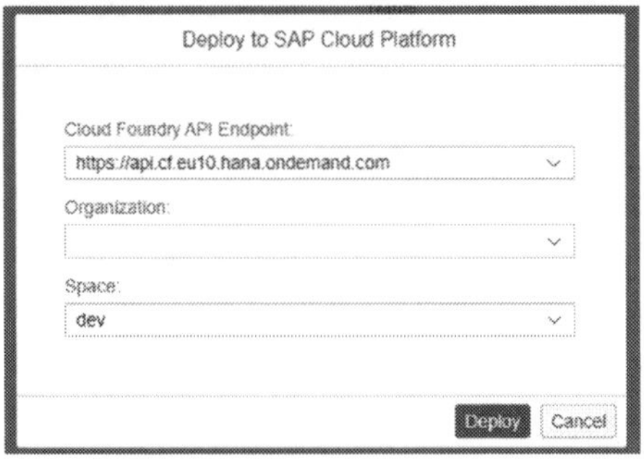

Figure 5-23. *Select organization and space in CF*

To validate the SAP Cloud Platform CF deployment, return to the Space and notice the number of applications and services.

To validate the deployment and the CF environment from the Linux terminal, run the following commands:

1) cf login -a <end_point_uri> t

    The Organization details are displayed if login is successful as shown in Figure 5-24.

```
                > cf login -a https://api.cf.eu10.hana.ondemand.com
API endpoint: https://api.cf.eu10.hana.ondemand.com

Email:

Password:
Authenticating...
OK

Targeted org

Targeted space dev


API endpoint:     https://api.cf.eu10.hana.ondemand.com (API version: 3.81.0)
User:
Org:
Space:            dev
```

Figure 5-24. *Connecting to CF endpoint using CF CLI*

2) cf services are shown in Figure 5-25.

```
                  > cf services
Getting services in org                / space dev as                ...

name                service     plan              bound apps   last operation     broker
                                            upgrade available
CROSS_SCHEMA_SVC    hanatrial   hdi-shared                     create succeeded   sm-hana-broker-
c6f315-7736-4e4b-bb4b-9d580ea8ec09
hdi_db              hanatrial   hdi-shared                     create succeeded   sm-hana-broker-
c6f315-7736-4e4b-bb4b-9d580ea8ec09
WAVEPRESS_UAA       xsuaa       application                    create succeeded   sm-xsuaa-9ef36:
4194-a399-54d4361e29b5
```

Figure 5-25. *Retrieve services inside the CF account*

Test out other commands from the CF/XS CLI if
other information needs to be retrieved from the
SAP Cloud Platform account. Available commands
can be found in the official SAP documentation site:

```
https://help.sap.com/viewer/4505d0bdaf494844
9b7f7379d24d0f0d/2.0.03/en-US/8bb90c8be91243
1dab74877a492a3f5a.html#loio8bb90c8be912431d
ab74877a492a3f5a__section_xkm_5k2_ys
```

Scaling microservices via SAP HANA cockpit for XSA

Scalability of software systems is one of the great features offered by cloud
environments. SAP HANA XSA being architected under CF principles also
supports such features. Scaling software systems means that an application
or service may have multiple instances of itself running at the same time
to meet the demand capacity of a software system. Scaling an application
or microservice will need to be done by both developers and XSA cockpit
administrators working together.

As applications grow and progressively get promoted from a development environment into quality systems and eventually reach production, some factors may degrade how software systems perform. Typically, a quality environment mimics the same company's production environment, and systems are more thoroughly tested in this environment, anticipating hiccups before reaching production. Often, the number of users and the volume of data increase in a production environment, resulting in application performance degradation. Other factors such as network bandwidth, business-related events, and perhaps additional unknown factors may present that were not encountered before. How can an application or service meet this demand? This is where scalability is applied to applications in XSA.

In the current exercise, an application and a service have been created as a single instance application. It can be seen from the SAP XSA cockpit in Figure 5-26.

◊ Space: development - Applications

All: 3

Requested State	Name	Instances	Disk Quota	Memory
Started	02RhD86I10w7WQ0Q-wavepress-ui	1/1	Unlimited	512 MB
Started	di-builder	1/1	Unlimited	256 MB
Started	hzooM4A49nLa4RsZ-wavepress-api	1/1	Unlimited	512 MB

Figure 5-26. *XSA applications seen from the XSA cockpit*

The same information can be read from the XS CLI using the **xs apps** command from the Linux terminal as displayed in Figure 5-27.

```
Found apps:

name                            requested state   instances   memory   disk
─────────────────────────────────────────────────────────────────────────────
82RhD86I18w7WQ8Q-wavepress-ui   STARTED           1/1         512 MB   <unlimi
di-builder                      STARTED           1/1         256 MB   <unlimi
hzooM4A49nLa4RsZ-wavepress-api  STARTED           1/1         512 MB   <unlimi
```

Figure 5-27. *XSA applications seen from the XS CLI*

Notice that there is only one instance running of one available instances in the system. If the application or service were to be stress tested, at some point, it would show a performance degradation if X number of requests were sent to the system within a period of time.

It makes total sense to increase the memory assigned to the application to meet such demand; however, a developer or SAP XSA cockpit admin can also create additional instances of the same application so that the requests coming in to the application or service can be load balanced into multiple instances rather than having a single instance processing all requests.

The question becomes when should an application increase its memory vs. when should an application have multiple instances running? The industry term for memory increase vs. instance increase is called horizontal and vertical scalability.

Horizontal scalability means that there will be more/less instances of an application. Scaling horizontally benefits the system to load balance requests and spread these requests equally to any of the available instances. The more instances running means that more resources are used, and therefore there is a price associated with this approach to keep active instances running. This approach may be required if the instances are placed in different geographical regions to ensure low latency when responding to application requests or if an anticipated event will occur and the business knows that it will expect a heavy load temporarily, as shown in Figure 5-28.

Figure 5-28. *Microservice instance representation*

Vertical scalability means that additional memory limits will increase/ decrease based on the need of the system. Vertical scalability will incur additional costs as the memory resources increase. Proper planning of scalability should be done to analyze the memory needs.

Based on the provided information, both scenarios are demonstrated as follows.

To increase the memory of an application (scale vertically)

1) From the SAP Web IDE, open the mta yaml file in the graphical editor.

2) Identify the application module that needs to increase memory limits.

3) Navigate to the parameters section and add memory property within the module and its desired value (Figure 5-29).

4) Save and rebuild the module.

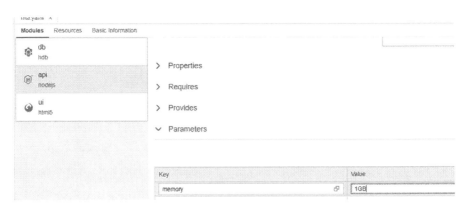

Figure 5-29. *Scaling microservices from the mta yaml file*

Validate from the console that the module was successfully built.

Requested State	Name	Instances	Disk Quota	Memory
Started	02RhD86I10w7WQ0Q-wavepress-ui	1/1	Unlimited	512 MB
Started	di-builder	1/1	Unlimited	256 MB
Started	hzooM4A49nLa4RsZ-wavepress-api	1/1	Unlimited	512 MB

Figure 5-30. *(Before) validation of scaling microservices as seen in Figure 5-30*

Restarting the application should take effect. Notice the service being restarted in Figure 5-31.

Requested State	Name	Instances	Disk Quota	Memory
Started	02RhD86I10w7WQ0Q-wavepress-ui	1/1	Unlimited	512 MB
Started	di-builder	1/1	Unlimited	256 MB
Started	hzooM4A49nLa4RsZ-wavepress-api	0/1	Unlimited	512 MB

Figure 5-31. *(After) validation of scaling microservices*

If the application restarts and it does not show the memory value updated, it is probably due to the values being overwritten by the system such as displayed in Figure 5-32.

Requested State	Name	Instances	Disk Quota	Memory
Started	02RhD86I10w7WQ0Q-wavepress-ui	1/1	Unlimited	512 MB
Started	di-builder	1/1	Unlimited	256 MB
Started	hzooM4A49nLa4RsZ-wavepress-api	1/1	Unlimited	512 MB

Figure 5-32. *Scaling microservices – application restart*

If the application memory does not assign memory value as instructed from the mta yaml file, proceed to execute the xs scale command from the Linux terminal as shown in Figure 5-33.

```
ost:hxeadm> xs scale hzooM4A49nLa4RsZ-wavepress-api -m 1GB

This will cause the app to be restaged and restarted.
keally scale the app "hzooM4A49nLa4RsZ-wavepress-api" with the new values? (y/n) > y
Jpdating app "hzooM4A49nLa4RsZ-wavepress-api" in org "l              ' / space "development" as XSA_ADMI
{...
}taging app "hzooM4A49nLa4RsZ-wavepress-api"...
```

Figure 5-33. *Scaling microservices from xs cli*

Return to the XSA cockpit and refresh the same screen that displays
the memory assigned to the applications to see and validate that the
application's memory was indeed scaled to 1GB (or 1024MB) from the
original 512MB memory as shown in Figure 5-34.

Requested State	Name	Instances	Disk Quota	Memory
Started	02RhD86i10w7WQOQ-wavepress-ui	1/1	Unlimited	512 MB
Started	di-builder	1/1	Unlimited	256 MB
Started	hzooM4A49nLa4RsZ-wavepress-api	1/1	Unlimited	1024 MB

Figure 5-34. *xsa cockpit after scaling applications*

To add additional instances of the same application (scale
horizontally)

1) From the SAP XSA cockpit, click the application
    name to drill into the application details screen.

2) Near the top of the screen, there are actions (Restart,
    Restage, Start, Stop, +Instance, –Instance, Delete) that
    can be performed to the application (Figure 5-35).

3) Select the "+ Instance" button to add an instance of
    the same microservice.

Figure 5-35. *XSA cockpit application instance operations*

> 4) The system immediately adds an instance as seen in
>   Figure 5-36.

Figure 5-36. *XSA cockpit adding instances to application*

> 5) After a few seconds, the system will display that
>   two of two instances are running once they become
>   available. The result of this operation shows that
>   two instances are running, and each of them has
>   1024MB (1GB) of memory, resulting in the service
>   (api module) being scalable in Figure 5-37.

6) Optional – Run additional requests against the endpoint to see if the service still runs (running stress testing in the system may showcase the comparison of one instance vs. two instances, and their response time would be helpful to see the scalability in action).

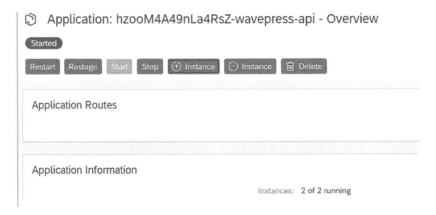

Figure 5-37. *XSA cockpit showing added instance running*

When multiple instances of an application or service exist, the request still is sent to the same URL : PORT, then the system internally knows how to check the load on each of the instances and smartly decide where to send the next request. This is the load balancing process.

Scalability does not mean that an application should only grow and be able to respond to stressful loads. Scalability also means that the microservice can be scaled down when resources are not needed (to save resources and money – as cloud services charge for the consumption of their resources/memory).

From the XS CLI (Linux terminal), proceed to scale the application down, back to 512MB, using the following command: ***xs scale*** <app_name> -m 512MB.

Returning to the XSA cockpit, validate the application (api module) has successfully been scaled down to 512MB. Notice that even though the application memory was scaled down, there are still two instances of the application running as seen in Figure 5-38.

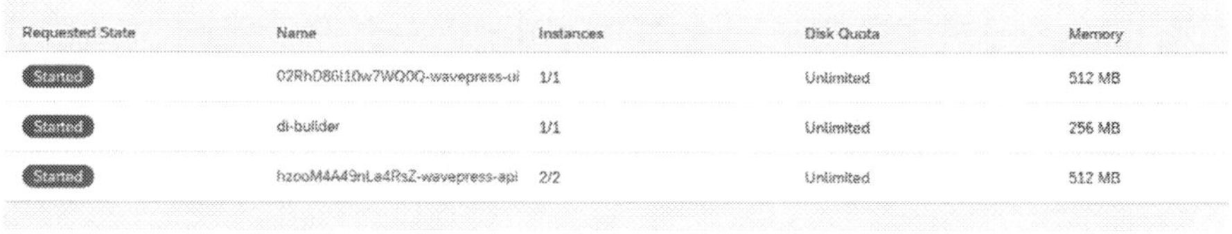

Requested State	Name	Instances	Disk Quota	Memory
Started	O2RhD8Gi10w7WQOQ-wavepress-ui	1/1	Unlimited	512 MB
Started	di-builder	1/1	Unlimited	256 MB
Started	hzooM4A49nLa4RsZ-wavepress-api	2/2	Unlimited	512 MB

Figure 5-38. *XSA cockpit showing 2 out of 2 applications running*

To remove one of the instances and set the application back to a single instance, perform the following steps (Figure 5-39):

1) In the SAP XSA cockpit, select the application.

2) Once inside the application details screen, notice the "– Instance" button and click it.

Figure 5-39. *Removing application instance*

3) The system will perform the action to destroy (remove) one of the instances. After a few seconds, it will display the new number of instances running.

4) Navigate back to the Space screen to show the applications running, displaying the number of instances and the memory values set.

🗆 Space: development - Applications

All: 3

Requested State	Name	Instances	Disk Quota	Memory
Started	02RhD86I10w7WQOQ-wavepress-ui	1/1	Unlimited	512 MB
Started	di-builder	1/1	Unlimited	256 MB
Started	h2ooM4A49nLa4RsZ-wavepress-api	1/1	Unlimited	512 MB

Figure 5-40. *XSA cockpit showing application instance after removing one*

Conclusion

Keep in mind that the scalability feature is very powerful when events occur and businesses need to be able to react to them to meet the consumer needs.

While many companies are figuring out ways to upgrade their SAP HANA systems and moving into the XS advanced architecture, this book explains the infrastructure and architectural changes involved that are required to start this process. All examples of this book interact with a SAP HANA database as the foundation where data is read and stored. Moreover, the book covers topics for developers trying to implement Node JS microservices in the SAP HANA XSA landscape.

Throughout the chapters, we have learned about the CF architecture, the UAA service implementing OAuth authorization, the JSON token generation and validation, and how it works in a SAP HANA XSA environment. The importance of understanding this architecture and its security is fundamental to building and deploying Node JS microservices running in XS advanced. After presenting the XSA architecture in comparison to Cloud Foundry's, the book dove into the development of several components from the SAP HANA database. Utilizing a classical schema table, design time objects such as a calculation view, synonyms, and a stored procedure were presented for the back-end side of the API. Consuming and exposing data as NodeJS microservices in an XSA environment were demonstrated as the focus point of the book. Finally, the deployment process of these microservices and the software versioning steps were shown. The book concluded with the scalability of applications and steps to deploy these microservices in a cloud environment.

APPENDIX A

Sources Used in Writing This Book

1) Cloud Foundry www.cloudfoundry.org/

2) SAP Developers https://developers.sap.com/

3) YAML https://yaml.org/

4) NPM www.npmjs.com/

5) SAP HANA www.sap.com/products/hana.html

6) REST https://restfulapi.net/

7) POSTMAN www.postman.com/

8) POST Request from POSTMAN https://blogs.
   sap.com/2018/07/16/testing-rest-apis-in-xsa-
   applications-without-ui-layer/

9) OAuth2 https://oauth.net/2/

10) JWT https://jwt.io/

11) SAP SDI https://help.sap.com/viewer/
    eb3777d5495d46c5b2fa773206bbfb46/2.0.01/en-
    US/ba66eb6311e345a6b88cd90afdfe84ac.html

© Sergio Guerrero 2020
S. Guerrero, *Microservices in SAP HANA XSA*, https://doi.org/10.1007/978-1-4842-6118-7

12) Flowgraph scheduling `https://help.sap.com/viewer/d60a5abb34d246cdb4ab7a4f6b9e3c93/2.0_SPS04/en-US/701be86d4ad3470485ee4ecb25258ba2.html`

13) SAP SLT `www.sap.com/products/landscape-replication-server.html`

14) GIT `https://git-scm.com/`

15) CORS/CSRF `https://help.sap.com/viewer/4505d0bdaf4948449b7f7379d24d0f0d/2.0.03/en-US/b8236393290048dda17b4545d17eac66.html`

16) CF CUPS `https://docs.cloudfoundry.org/devguide/services/user-provided.html`

17) The SAP Web IDE `https://developers.sap.com/topics/sap-webide.html`

18) GitHub Help `https://help.github.com/en/github/getting-started-with-github/create-a-repo`

19) SAP HANA XSA, Access a classic schema from an HDI container `https://developers.sap.com/tutorials/xsa-create-user-provided-anonymous-service.html`

20) Passport JS `www.passportjs.org/`

21) Body-parser `www.npmjs.com/package/body-parser`

22) Express JS `http://expressjs.com/`

23) Application Router Configuration `https://help.sap.com/viewer/4505d0bdaf4948449b7f7379d24d0f0d/2.0.03/en-US/5f77e58ec01b46f6b64ee1e2afe3ead7.html`

24) Managed vs. existing service `https://help.sap.com/viewer/4505d0bdaf4948449b7f7379d24d0f0d/2.0.04/en-US/4050fee4c469498ebc31b10f2ae15ff2.html`

Index

A, B

Application program interface (API)
    REST APIs (*see* Representational
        State Transfer (REST APIs))
    request *vs*. XS engine, 9
    SAP HANA XSJS, 7–8

C

Cloud Foundry (CF)
    applications/services, 11
    client request/endpoint, 13
    cloud/XSA architecture, 14
    command-line interface, 14–15
    dynamic environment, 15
    express, 17
    Hello World program, 16–17
    hierarchical organization/
        concepts, 10–11
    microservices, 18
    Node JS, 15–16
    node program creation, 16
    npm pricing section, 17
    organization hierarchy, 11
    resource endpoint, 13
    roles, 11
    router/UAA service, 12
    Web IDE environment, 84

Cockpit (SAP HANA XSA)
    controller, 107
    custom services, 110
    host management, 115
    landing screen/page, 106–117
    marketplace/instances, 109
    migrating option, 117
    monitoring, 109
    navigation bar, 106–107
    organization, 108
    pinned hosts, 112
    role collections, 113
    routes, 111
    service connection details, 111
    spaces, 107–109, 112
    tenant databases, 115
    trust
        certificate, 115
        configuration, 113–114
    user creation/management, 116
    user-provided services, 110
Command-line interface (CLI), 14
Compatibility mode, 134–136
Continuous integration and
        continuous delivery
        (CI/CD), 27
Cross-origin request sharing
        (CORS), 42

© Sergio Guerrero 2020
S. Guerrero, *Microservices in SAP HANA XSA*, https://doi.org/10.1007/978-1-4842-6118-7

Made in the USA
Middletown, DE
10 December 2021